WORKING
THE TIDES

WORKING
THE TIDES

A Portrait of Canada's West Coast Fishery

Peter A. Robson
and Michael Skog, editors

HARBOUR PUBLISHING

Published by
HARBOUR PUBLISHING
P.O. Box 219
Madeira Park, BC Canada V0N 2H0

Published with the assistance of the Canada Council and the Government of British Columbia, Cultural Services Branch.

Cover design by Roger Handling / Terra Firma
Text design by Mary White
Cover photograph by Vance Hanna
Endsheet photograph by Alexandra Morton
Paintings by Graham Wragg
Line drawings by Alistair Anderson
Printed and bound in Canada

Canadian Cataloguing in Publication Data

Main entry under title:

Working the tides

Includes index.
ISBN 1-55017-153-4

1. Fisheries—British Columbia. I. Robson, Peter A. (Peter Andrew), 1956– II. Skog, Michael, 1967–
SH224.B7W67 1996   338.3'727'09711 C96-910490-1

CONTENTS

SEINING FOR SALMON

SEINING FOR HERRING

GROUNDLINE FISHERIES

TRAWL FISHERIES

JIGGING AND TROLLING
FOR ROCKFISH

DIVE FISHERIES

AFTERWORD

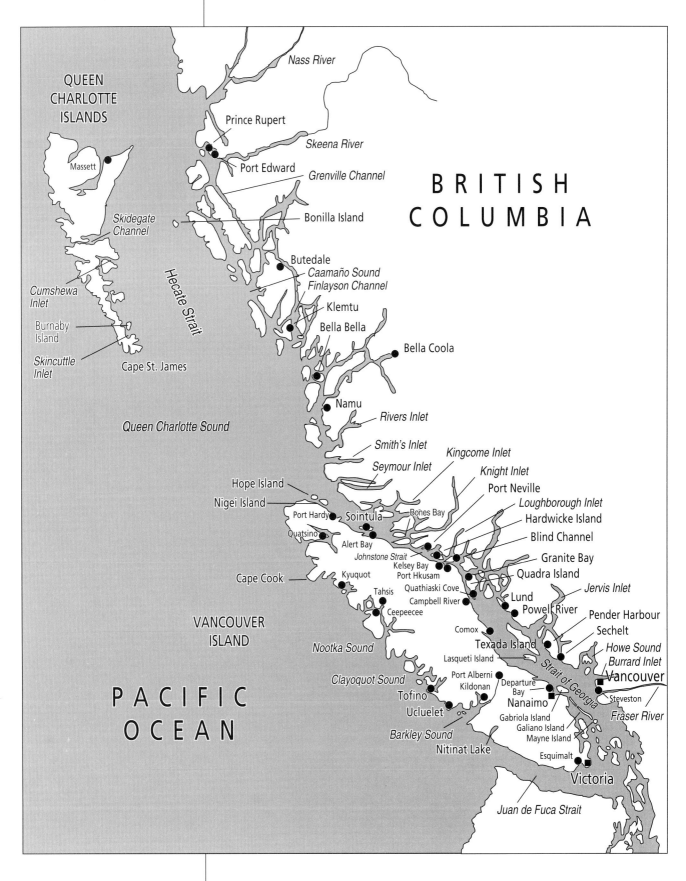

QUEEN
CHARLOTTE
ISLANDS

Massett

BRITISH
COLUMBIA

Nass River

Prince Rupert

Port Edward

Skeena River

Grenville Channel

Bonilla Island

Butedale
Caamaño Sound
Finlayson Channel

Klemtu

Bella Bella

Bella Coola

Skidegate
Channel

Hecate Strait

Cumshewa
Inlet

Burnaby
Island

Skincuttle
Inlet

Cape St. James

Namu

Rivers Inlet

Queen Charlotte Sound

Smith's Inlet

Seymour Inlet

Kingcome Inlet

Knight Inlet

Port Neville

Loughborough Inlet

Hardwicke Island

Blind Channel

Hope Island

Nigei Island

Port Hardy

Sointula

Boñes Bay

Granite Bay

Quadra Island

Jervis Inlet

Quatsino

Alert Bay

Johnstone Strait

Kelsey Bay
Port Hkusam

Quathiaski Cove

Campbell River

Lund

Powell River

Pender Harbour

Sechelt

Cape Cook

Kyuquot

Tahsis

Ceepeecee

Comox

Texada Island

Howe Sound
Burrard Inlet

Vancouver

VANCOUVER
ISLAND

Nootka Sound

Lasqueti Island

Strait of Georgia

Steveston

Clayoquot Sound

Port Alberni
Kildonan

Departure
Bay

Tofino

Nanaimo

Fraser River

PACIFIC
OCEAN

Ucluelet

Barkley Sound

Nitinat Lake

Gabriola Island
Galiano Island
Mayne Island

Esquimalt

Victoria

Juan de Fuca Strait

PREFACE

Working the Tides

Who among us hasn't strolled along a wharf and gazed upon rust-streaked fishboats and what looks like a tangle of rigging—and wondered, perhaps wistfully, just what that life would be like. What is it about the fishing life that has drawn us to it since the beginning of human history?

Fishers, by nature, have never been great seekers of glory. Generations have quietly gone about their business far from the eyes of the general public. Apart from some rare accounts by working fishers or their admiring spouses, reporting about British Columbia's fishing industry has been produced at a desk, or occasionally on a wharf. In the trade press, stories generally focussed either on business or political news. Photos, when there were any, showed smiling executives in suits or union organizers addressing groups on docks. The only boats that rated mention were typically shiny new ones with decks that had never run red with fish blood.

In 1986 all that changed. It was then that a bright young entrepreneur named Leon La Couvée, who had grown up in the fishing community of Ucluelet, started a magazine called the *Westcoast Fisherman*. It was the inspiration of La Couvée and his first editor, Richard Gross, to write about the industry from the other side up, giving centre stage to the boats and people who actually did the fishing. From the first, the magazine was a group effort which benefitted from a succession of editors including Gross, Alan Haig-Brown, Terry Clarke, myself, Michael Skog, and the current editor, David Rahn, who has been with the magazine longer than anyone else.

Once the forum for this inside-the-skin type of writing was created, a community of writers came forth, eager to offer material. Thanks to their raw enthusiasm, they often found themselves on board working fish boats with only camera, notebook and sleeping bag, documenting the experience by passing wrenches in the engine room, spelling a crew member at the wheel, making coffee or doing dishes. Sometimes getting a story meant relaxing around the galley table with a fishing crew and a pot of coffee, poking around boatyards, sitting in a fisher's home looking at scrapbooks, or clinging to a deck in a raging storm in Queen Charlotte Sound. This fresh new style of writing about the fishing life was rewarded with strong support from the industry. The fish boat that didn't have the current issue handy was rare indeed.

Many of those who wrote for the magazine were inexperienced. They were seeking a chance to break into the field of writing—or, like me, the chance to have an adventure or two, and to witness a life to which few are given access. Some writers were working fishers; some had never been aboard a fish boat. Many got seasick. Many, like Alan Haig-Brown, Brian Gauvin, Vickie Jensen, Robert Morris, Jean Rysstad, Joan Skogan and I, survived to write books of our own.

All the writers were part of a journalistic flowering that made an essential contribution to documenting the lives of British Columbia's working fishers. Whether the writing was smoothly professional or as rough and craggy as the west coast itself, the underlying spirit of it was as real as the fish in the hold—and it still is.

For me, it was Alan Haig-Brown who embodied the spirit that the *Westcoast Fisherman* became famous for. As the magazine's second editor and then as executive editor of Westcoast Publishing's expanding fleet of magazines, he brought with him a deep familiarity with and a great respect for the men and women who got fish scales on their gumboots.

All but three of the pieces in this book were originally published in some form by the *Westcoast Fisherman*. Gathered together, they present a unique overview of British Columbia's commercial fishing industry that will surprise some readers. If we were to judge by accounts in the media, we would think the only fishery on the west coast is the salmon fishery; and that there are only two things to say about it: it is on its last legs, or it is enjoying a rich bonanza. This book is a reminder that over eighty different species are harvested on the BC coast, involving 20,000 workers whose collective experience forms a rich historical tradition larger than the cyclical ups and downs of any one fishery.

In 1996 the *Westcoast Fisherman* celebrated its 10th anniversary. We salute all those who helped the magazine reach that milestone.

We invite you to meet our west coast fishers, to travel with them to out-of-the-way places known only by them, to share their hopes and fears, and perhaps to gain some insight into their unique culture. These are the men and women who define British Columbia's commercial fishing industry—those who make a living "Working the Tides."

PETER A. ROBSON

A Fishing Legacy

Since our beginnings in 1986, we at the *Westcoast Fisherman* have had the privilege of documenting the lives and adventures of the people who live and work in the fishing communities of British Columbia. These communities celebrate the very best of the human spirit—hard work, courage, integrity and intelligence.

The west coast, with its wild coastline, complex waterways and unpredictable weather, can be a cruel master or a willing partner. In either incarnation, it is accepted and nurtured by the fishermen who bring forth its bounty from the sea—a bounty that with good husbandry will always be available to us.

Please enjoy the writings that follow. They are an exceptionally precious piece of our British Columbia heritage.

LEON E. LA COUVÉE, PRESIDENT
DAVID RAHN, GROUP PUBLISHER

TROLLING
FOR
SALMON

Previous page: A troller at work south of Alert Bay. Photo: Peter A. Robson.

Strait of Georgia

HANDLINES AND TORTURE STICKS
Dave Miller Recalls

Rob Morris

Native fishermen were handlining or trolling from their dugout canoes long before Europeans arrived on the west coast. The natives began to commercially fish British Columbia's north coast and Queen Charlotte Islands in the late 1800s. They were soon joined by white fishermen in rowboats and skiffs. A 1917 fisheries department report estimated 500 troll fishermen on the north coast.

Handlining from dugouts and rowboats reached its peak during the depression years of the 1930s. Some 500 to 700 boats were working the Strait of Georgia area alone. The fishery continued into the 1940s, when gasoline-powered boats with mechanical gurdies replaced handlining as a viable fishery.

There are men and women on the coast today who took part in the handline fishery in those latter years. One of them was Dave Miller.

In the mid-1930s Dave Miller was fifteen. He ran away from school in Victoria with a couple of buddies and they headed up the Strait of Georgia to try their hand at commercially fishing "bluebacks" —coho salmon.

"I had seven lousy bucks in my pocket and an old dugout canoe. We had a tiny little gas stove, a Primus, but no tent, no nothin'. Geez, it was ragged, you know. So we got up to Hornby Island—we didn't know the

Handliner and a typical double-ended rowboat with a herring rake and hand-shaped dipnet. Photo: Hubert Evans.

HANDLINES AND TORTURE STICKS

Dave Miller's cod troller *Memories*. Miller started cod trolling after the Second World War and a stint in the "Gumboot Navy." Photo: Tor Miller collection.

score—and a guy says 'There's some shacks there on Toby Rock. Help your-selves.' The shack was all right; it had an oil drum for a stove, that's all."

Toby Rock—called Toby Island on today's chart—off the southern tip of Hornby Island, was a small islet, no trees and a half dozen shacks. Each season, the shacks were first come, first serve. At low tide one could cross the intertidal zone to a well on Hornby Island. In nearby Tribune Bay, a German farmer was more than willing to trade a few coho for fresh fruit.

The season opened May 1. The collector boat—either the *Tadeoshi* or the *Cape Sun*—came out in the morning and evening every day to pick up fish. When the Japanese fellow running the collectors learned Miller and his pals were from Victoria and were trying to catch some bluebacks, he set them straight. "Gear you got no good; you catch nothing with that." So he rigged them out with the popular Coho King spinners and showed them how to cut the "chunk" out of a salmon throat, slice a "V" in it and hook it behind the spinner. With about eight fathoms of gear out, his instruc-tions were "Troll slow, and turn, turn, turn!"

"He was a pretty good guy—Toshe Yuedi was his name—from Victoria," recalls Miller. "He really knew what the score was and I guess he took pity on us."

Miller recalls that the Japanese buyer was paying 7 cents a pound for dressed fish, 6 cents in the round, overall a cent more than the white buyers. The collectors delivered to larger packers, such as the 45-foot *Lake Beewah* in Deep Bay, which had ice on board to deliver fish to Vancouver.

The mounted police had taken Miller's fishing partner back to his parents, leaving him alone. "I stayed at Toby that part of the summer and then the fish died off, and I see a lot of the guys leaving. 'We're going to Cape Mudge.' 'Where the hell's Cape Mudge?' I asked. I didn't have a bloody clue!

Another of Miller's cod trollers, *Swan Point*. Photo: Dave Miller collection.

Dave Miller at his home in Victoria, 1993. Photo: Rob Morris.

"So I headed to the Cape too. There I was in that open canoe. Geez, if the wind had come up...The packer stopped me off Cape Lazo.

"'You OK, kid?'

"'Yah, I'm goin' to the Cape.'

"He pointed. 'It's way up there.'

"'Holy geez,' I says, 'and I've just got torture sticks'—that's what we called oars."

When Miller got to Cape Mudge, he found more than a hundred driftwood shacks on the beach. He moved into one on Dogfish Bay, which was around from the Cape, facing Cortes Island. "It was an awful shack, but at least it had a bunk and the roof didn't leak! I was in there for a month! We fished the inner kelp, then the outer kelp [about a quarter mile and a mile offshore, respectively]."

I remember the guy on the collector, the *Double Island*, asked me how well I made out one day. Well, I said, I don't know, I got 25 fish. He said, 'That's not too bad for a greenhorn. The high boat here was fifty.' Geez, there was a hell of lot of fish there. Well, actually Cape Mudge was *the* best place to fish. Some guys stayed there all year, others moved around. I was one that moved around."

On board the collector boats were scales for weighing the fish, a couple of 45-gallon barrels of gas for the small gas trollers, but no fresh water until water delivery out to the grounds was won as a concession after the 1935 blueback strike. Their stern cabins, which were normally used as an accommodation, were stocked with groceries and fishing gear.

"The grub was cheap, you know. You'd go and dump off your fish and he'd give you a slip and tell you to go back to the stern cabin. You'd take this and that, then peel some money off and give it right back to him—he was the fish buyer and the storekeeper! On the slips he used to put 'Victoria Boy,' not D. Miller. I was the only one that came from Victoria. The rest came from, God knows, all over the place."

Miller stayed at the Cape all summer, then moved back down to Hornby Island. The *Double Island* came in saying the season was over, it was the middle of September, and the last packer run to Vancouver. He offered Miller a tow to Nanaimo. "He asked me how much money I made. I looked in my purse, I had fifty-five bucks for the whole bloody summer!" 🐟

GUBBY GUDBRANSEN
Ucluelet Day Troller

story and photos by Terry Clarke

Ucluelet

Taking a look at some of the photos of Al "Gubby" Gudbransen's earlier boats, you can see he settled on a favourite day troller design a long time ago. All of his day trollers save one, including the *Seabird IV* which he was running out of Ucluelet in 1988, have been double-enders with a round wheelhouse and no bunk to speak of, and no holds—just three checkers. All were built between 1934 and 1941 and ranged in length from 33 to 37 feet, with the narrow beam and deep draught which still give the classic day troller design its character. The *Seabird IV* was built at the famous Kishi Boatworks in Steveston. In 1937, new, she cost $1,200.

Earlier in the day, we'd tacked back and forth along a run

"I've been at this racket a lot of years and I haven't got rich yet, so I don't expect it to happen in the next few years."

The *Seabird IV*, built for $1,200 at the Kishi Boatworks in 1937.

Gudbransen says is called "Gubby's tack," looking for the coho which had been there in good numbers the day before. Then he'd stayed out late, coming back just before 2200 hours with $1,300 worth of fish and a cold supper for his efforts. Today, as a voice over the radio put it, "It's no use out here. There's no feed or bugger all." Gudbransen shakes his head in agreement, saying, "You don't seem to get two days in a row, I don't know why. It doesn't take many fish to make a dollar, though," he says in reference to 1988's higher prices on coho and springs.

Later on, as the day turned out to be as quiet as the weather, Gudbransen said, "I guarantee you that if we still had a six-month season, 90 per cent of these boats would be in right now. The only reason they're dragging around here now is the price."

Gudbransen had started the day at 0430 hours, moving out past the lighthouse at Amphitrite Point before 0500. By 1030 hours he had only eight coho, then things picked up just a bit. "That's what I like to hear, the fast ring of a coho bell. That's the way they're supposed to come—four and five to a line. And they're starting to get a bit bigger. That's a good sign." It turned out to be misleading, unfortunately, with only thirteen more coho taken by 1330 hours. Gubby kept tacking, always within sight of Amphitrite Point, no longer expecting another big day and only half seriously wishing he had stayed in seining for more than the one season back in 1967.

"For this racket, you've got to have lots of patience," said Gubby. "This is where I can make it and the big boats can't. They need good production to meet their costs. I can keep scratching and make a go of it. A few more passes, another dozen coho and a couple of smileys [large chinook, or spring, salmon], and that will be something anyway."

Gudbransen listened to the radio for news of activity elsewhere, but he wasn't worrying over it. "If they're catching some, good for them. I can't spend what they've made, and I can't catch what they've caught."

Day trollers are a different breed than the ice and freezer trollers, says Gudbransen. "The freezer and ice boats are the ones who push, because they have to," and their operating costs are much higher. "Day trollers, they like a dollar if it comes, but they're not going to bust their ass to get it, the guys I know anyway. Sure they get frustrated, but all they want is a comfortable living. I've been at this racket a lot of years and I haven't got rich yet, so I don't expect it to happen in the next few years. I hear so many people, and all they think about is money. Jesus, it gets to me."

Morning evaporated into a hot August afternoon. The sun made us squint as it glinted off the ground swells. "I hate August," said Gubby. "Nine times out of ten it's an evening bite or a late bite. There used to be a saying around here that if the fish don't bite by two, you may as well go home. It's not so much like that any more. These hatchery fish are a different breed."

There were no more coho or chinook caught in the afternoon to help make a slow day decent. "That was a stroke of luck," said Gudbransen, thinking about the 30-pound smiley earlier in the day. "I used to do really well on springs, but I seem to have lost my touch a bit. There aren't many to target on any more."

Despite the chance of another evening bite, Gudbransen chose to be home in time for dinner. "I've fished so many years now, that when I'm ready to go in, I say one more tack and whatever I get, that pays for fuel." At the end of the day, Gubby delivered 43 coho, earning himself $500 for the day.

There aren't many more of those last runs to be made in the *Seabird IV*, Gudbransen explained over a sandwich at sea in the morning. "I've got this one sold, so I'm going to have one built in October, at Albion, one of the original Albions with the round, trolling hull." Gudbransen explained that to make the changes he'd like to see on the *Seabird* would probably cost him $40,000, "because you don't know what you're going to get into once you start working on an old boat like this. For an extra $70,000 I can get a new boat, so why not?"

Gubby sets a knot on one of his leaders with the help of a little spit. "Another reason I'm going ahead with the new boat, though, is because one time my son, who's twelve now, said to me, 'Dad, if something happens to you, can I have the *Seabird IV*?' And I said sure. He's crazy on fishing, so is my grandson who is five, but in this age a young guy can't buy into fishing. When I bought my first boat, the *Marine K*, it was $4,500 and an average season was $5,000. He needs help, so I'm going to give my son the opportunity anyway. I'm hoping they'll take the boat over, so I can direct traffic from the radiophone." 🐟

Gubby sets a knot, with the help of a little spit.

Cleaning fish in the checkers.

Illustration by Graham Wragg

ELMER AND THE KID

a short story by Pete Fletcher

Elmer knew about the engine; knew about it from the day he first laid eyes on it—so he says.

It was sullenly secured in the belly of his soon-to-be boat, imprisoned to the keel and ribs with rusty bolts. Habitually neglected by past owners, it had seen an oil change about as often as some of the skippers had changed their underwear. She was forlornly coated in rust flakes. Maintenance had remained a distant word to be found somewhere in the archives of a promised land—in a dictionary under "M."

Like many things denied the sustenance levels of good health, the engine developed a consumptive cough, loose joints, shook with advanced palsy, and now was petulantly temperamental—suitable for its role as the chronic invalid. "If man made it, man can fix it," said Elmer. He bought the boat with this air of confidence. The former owner, thumbing through a thick wad of bills, made haste off the dock with never a backward glance.

Elmer's newly acquired troller, despite its unwilling propulsion unit, had at least been outfitted in a more generous hand in the upkeep of the outside working gear. And after all, as Elmer said over and over to his deckhand, "Fish in the hold—money in the pocket. That's what it's all about." Somewhere he figured that "goldamned gas engine" could wait until his self-imposed list of priorities was whittled down.

Well it never really did get shaved to the size he wanted, and Elmer and "The Kid" sailed, ready-or-not, when the season began.

"The Kid" was seventeen years of age and had never been fishing. Still, he was game enough, could cook passably fair, and never got seasick. Elmer was willing to forgive his lack of experience and tender youth in light of these other unexpected attributes.

There may have been something about the sound of the intestinal troubles of the diseased engine—maybe it was the vibrations caused by the on-going seizures of shuddering which overtook it—or perhaps it was just the overall mechanical melancholy resonating throughout the hull—but the sea lions loved the little troller. Elmer soon established a loyal following wherever he went. Understandably, this made fishing a tough proposition and sharing just wasn't in Elmer's nature. Soon the battle lines were drawn. Half chewed fish, moon-shaped bites, and severed leaders became commonplace, and to his credit, Elmer fought back with every trick he knew. Through it all, however, his faithful herd of mammalian salmon-lovers stayed with him. To The Kid, mornings were the worst. Elmer would descend into the tiny trunk cabin and begin the morning routine.

"Mebbe it's the spark." The muttered words would assume the air of an ancient chant. "Mebbe it's the goldamn carburetor." Religiously, as he ran through the checks, a steady drone carried the snatches of—"Mebbe this connection's loose." An hour or so later, the fitful sounds of a tortured piece of machinery would begin to thump and Elmer would yell, "Up anchor Kid," and wobbly like—away they'd go.

At noon on a particularly stressful day, the exhaust manifold blew a hole through its flimsy skin. The Kid, cooking pork chops, was suddenly enveloped in a black shroud of noxious gases. Abandoning his frying pan, he fled to the safety of sea breezes and yelled to Elmer that the exhaust elbow had blown.

Elmer cut a piece of rusty stove pipe into a patch and copper-wired it to the offending structure, creating a metal bandage, which in time became a part of the regular features of the exhaust outlet. The Kid was going to chuck the pork chops, but Elmer allowed as how with ketchup, "They ain't half bad." After the repair, the climate in the trunk cabin and wheelhouse dropped from totally life-threatening, to one in which the hapless occupant had only to endure a constant burning of the eyes and a mild constriction of the throat. Elmer stocked up on a case of ketchup at the next fish camp they hit.

Fishing close to a big kelp bed on a day that was "Sure to make our season," Elmer was beside himself with self-congratulatory joy. No longer could be heard the familiar "Urk, urk, urk" a few hundred feet behind the boat. Gone was the entourage of whiskered and thrice-cursed sea lions. The kelp bed was host to a run of salmon—killing themselves to get on the hooks.

A rocky shoreline lay behind the kelp and a large southwest swell pounded with impressive strength against the bluffs, creating a din over which Elmer and The Kid had to shout to make themselves understood.

Ah! How the Gods smiled. An angelic beam lit up Elmer's battered old face. A choir of ethereal beauty serenaded them from jingling brass bells above.

And then the engine quit.

While The Kid tried to haul the gear in by hand, Elmer dove for the cabin.

Over the noise of the thundering surf wafted faintly the commonplace and not-so-reassuring sounds of "Mebbe there's water in the line"—"Mebbe it's these damn plugs, I've got old spares some'ere's."

Having dragged in the lines, which now lay in untidy heaps on the afterdeck, The Kid nervously said through the wheelhouse door, "Elmer—we're drifting through the kelp toward the shore."

"No spark! Lookit that. I gotta get me a diesel." Taking the distributor cap off, Elmer peered at the inside and said there was moisture there. Rubbing at the miscreant cap with the sleeve of his grey wool Stanfields, he yelled to no one in particular, "The damn thing's cracked."

Out on the deck, The Kid said "...Elmer, the rocks."

"Well, drop the bloody anchor then."

Fussing through some drawers, Elmer came out with a tube of glue and smeared a generous amount on the cap's insides. He opened a big box of kitchen matches, lit some up and held them under the cap to speed the drying. A burst of flame now and then told him they were too close. Lost in the depths of pyromania, he dimly heard The Kid on the bow.

"Elmer, the anchor's not holding."

"Well let some more line out."

"...The rocks."

"Mebbe it's the points"—rasping sounds of scratching sandpaper issued through the open porthole of the trunk cabin.

Over the deafening roar of surf, The Kid's anxious gaze told him they were only a few feet from disaster. The swells were cresting almost under their keel. The round-bottomed troller rolled like a drunkard.

"...Elmer!"

A spluttering cough of the engine was followed by an uneven rhythm. Elmer jammed her into gear, spun the wheel and pivoted on a cresting wave. "I've got to get me a diesel," he said.

The Kid hauled in the anchor and they chewed their way through the foaming kelp.

Several years passed. The Kid grew into a fine young man, and to the surprise of many, stayed in the fishing industry. It so happened one day, while walking down a line of floats, he spied the unforgettable shape of Elmer's boat. Thinking he'd haul up and relive old memories for a moment, he approached the vessel to hear muffled sounds coming from below. "Mebbe it's the voltage regulator"; this was followed by the bash of metallic blows. "I've got a spare here some'ere's."

Moving along briskly, The Kid recalled an appointment he was surely late for.

CHALLENGER
The Life and Times of Johnny Madokoro

Kenneth N. Howes, photos courtesy Madokoro collection

I n the winter of 1922, two gillnet boats, both of them wooden double-enders, travelled from Steveston on the Fraser River to a remote village on the rugged west coast of Vancouver Island. The trip was long—bucking tides with 5-horse Easthopes—but the weather was favourable. The stretch of coastline they navigated claimed more shipwrecks than perhaps anywhere else on the west coast of North America, yet their journey was without incident.

On board one of those boats a young boy, Yoshio Madokoro, caught the first glimpse of the coast that would become his home and the source of his livelihood.

There is a cold southeast wind coming up the Alberni Inlet, and a light chop developing on the water. The *Lady Rose* and the *Frances Barkley* bob

Port Alberni

Johnny with his troller at Port Alberni.

idly in the wake of a passing tug and the sky threatens rain. It's March 15, 1995 and apart from the sound of acetylene welders and the clank of steel at the ways, things are quiet in Port Alberni harbour. An Oldsmobile pulls into the parking lot near the harbourmaster's office. A figure appears from the vehicle and starts toward me. It is Yoshio Madokoro, or Johnny, as most people know him.

He remembers me from my days deckhanding with my father in Tofino. I remember him as an enigmatic veteran of the seas who would appear for a meal at the Maquinna Hotel, and the next day disappear on a long west–northwest tack up to Kyuquot. He had explained over the phone that he is usually at the golf course until the early afternoon, then he comes by the boat to "putter."

"Too wet," says Johnny. "What a winter," he adds in reference to the incessant precipitation that has hampered his time on the green. Johnny Madokoro likes to golf as often as possible. He always plays a full eighteen holes, and he does not use a cart. "My legs are still good," he says as he slaps his palm against his right thigh.

We take a table at the Blue Door Cafe, a nostalgic little greasy spoon just a hoochie's throw from the Port Alberni waterfront. As the lunch crowd thins, Johnny begins to speak about growing up around the bustling canneries of Steveston, where he was born in 1913. Johnny's father came to BC in the early 1900s from Wakayama-ken on the island of Honshu. He got a job as caretaker of a property on Gambier Island owned by a wealthy Vancouver businessman. Later he moved to Steveston, where Japanese fishermen were already established in the industry, and bought a gillnetter.

The decision to move to Tofino was probably prompted by mounting racial tensions between European and Japanese fishermen on the Fraser River, but also by reports of the good fishing and a secure harbour on the west coast of the island.

By 1922, Ucluelet had a thriving fishing community consisting of some fifty to sixty boats. Thirty miles away, in Tofino, there was a small settlement which included families with names such as Arnet, Erickson, and MacLeod. There was also the beginning of a commercial fishing fleet which Johnny Madokoro would eventually join. Johnny recalled the first commercial troller to fish out of Tofino, Iso Zaki, who claimed to be able to "catch a spring salmon with a tablespoon."

In those days, Tofino had a one-room school—grades one to eight—with one teacher. There was one doctor and a general store run by Tower & Mitchell. Provisions came in on the CPR steamship *Princess Maquinna* every ten days.

Johnny completed grade eight in the schoolhouse and, at his father's request, moved to Cumberland on the east coast of Vancouver Island to study Japanese. But during his second year in Cumberland Johnny's father passed away. As he was the oldest son, the burden of providing for his family fell on Johnny's shoulders and so, at age fourteen, he began his career in fishing. He started out as a deckhand with his uncle, and soon learned about commercial salmon fishing. They fished pretty near

Johnny Madokoro with a steelhead caught at Port Alberni in 1953.

everything back in those days: pilchard, herring, cod and dogfish as well as salmon.

"No gurdies back then," says Johnny with a grimace as he mimes the strenuous job of pulling in a trolling line by hand. "That was hard work." Eventually, the hard work paid off and somehow Johnny managed to support the family. The fishing season was long, but even in the winter there was rarely time for relaxation. "We used to go beachcombing and occasionally handlogging for firewood. Once we had the wood all chopped, we had to carry it up to the house, one load at a time." Johnny leans back casually in his chair as he thinks about all that back-breaking work. I wonder if he is thinking back with pride, or just simply glad that he doesn't have to lug twenty cords of wood up from the beach just to keep warm in the winter.

The 1930s were a busy time in and around Tofino. Pilchard reduction plants established themselves in Nootka Sound, Esperanza Inlet, and Tofino Inlet. American trollers from the Columbia River area fished off the coast of Estevan Point (there was no 200-mile exclusion zone before the Second World War), and the fishing season began in earnest after New Year's Day.

Johnny's boat on the left and his brother Thomas's on the right, at Storm Bay, Tofino in 1936.

"We started off of Hot Springs Cove fishing cod for 5 cents a pound. We had to save the liver because that was the most valuable part, for the oil. We got about 12 to 14 cents a pound for springs."

Around 1936 the Tofino Fishermen's Co-op, an organization of thirty or so members, formed as a way to gain autonomy from BC Packers, who had a lion's share of the salmon processing industry. Johnny Madokoro was treasurer of the co-op just before the Second World War, and he remembers how he had to sell the assets and distribute the funds between the members when the co-op disbanded.

The war had a great effect on the lives of those who lived and worked on the west coast of Vancouver Island. Many young men left the villages to enlist while those who stayed behind became part of the notorious "Gumboot Navy," the commercial fishermen and merchant marines who were the ostensible home guard. For the Japanese, however, war carried different dimensions of separation and hardship.

In 1942 Johnny Madokoro was one of thirty fishermen who were ordered to follow a government escort boat to New Westminster, where his troller was tied up with hundreds of other Japanese-owned fishing boats. Johnny returned to his home near Tofino—Storm Bay, as he knew it—and waited pensively for the inevitable. It was not long before the Navy sent word that the Japanese had twenty-four hours to leave their homes.

Johnny was sent to an internment camp in southern Ontario. He and about a hundred others from BC found themselves working as farm labourers, picking sugar beets, cucumbers, tomatoes and tobacco for 25 cents an hour until they could return to their homes and to their lives. In a photograph taken in 1943, Johnny pitches hay into a wagon, lean and resolute, wearing dungarees and looking at home on the farm.

Many of the men who were sent back east stayed there, Johnny among them. They found work similar to that which they had done during the war, but for better wages. But in 1952, Johnny found himself another troller and once again began to fish the west coast. Of course things had changed. The pilchard was gone and there was a road to Tofino, but what remained were those unchangeable, soul-sustaining elements of west coast life: the cloud-capped crest of Lone Cone on Meares Island, sunsets on an endless, unbroken western horizon, and the art and craft of catching salmon.

After a while, Johnny and I walk down to his boat, a double-ender aptly named *Challenger II*. There is a partially overhauled gurdy on the hatch and an assortment of tools around the deck. Johnny has been reclaiming some spare parts from an old gurdy someone gave to him earlier in the winter. Johnny bought the boat back in 1954, just after he returned from Ontario. He has fished her every season for the last forty years.

"And still haywire," Johnny says with a laugh.

Later, Johnny makes a pit stop at the local grocer's where he buys himself a ripe bunch of Costa Rican bananas. "I get hungry when I'm out golfing, so I take a banana with me to munch on," he explains. The clouds are supposed to clear tomorrow. Johnny plans on playing a round in the morning, then he will be at work down on the boat in the afternoon.

I ask Johnny if he has any particular secrets for longevity. He laughs and shakes his head. Personally I suspect it has something to do with a lifetime of hard, honest work, and a good sense of humour. A banana munched somewhere on the back nine might not hurt either.

Illustration by Alistair Anderson

COMMUNICATIONS CONUNDRUM

a short story by Pete Fletcher

"Yeah, it looks as if the weather here's going to be okay. Wind's likely to pick up though."

Through the electronic marvels of diodes, transistors, unified with solid state circuitry, driven by the magic medium of VHF, this innocuous piece of information sped through the atmosphere to Oscar's attentive ears. He replied:

"Right—well—all right. I'm going to pick up my gear and tidy up some. Maybe see you later."

Oscar broke all his previous speed records in yanking up lines and hardware. Punching the throttle down hard, he raced to join his friend Jerry—who was obviously killing salmon by the ton at their previously agreed-upon co-ordinates.

Such was the cloak of Jerry and Oscar's seemingly routine small talk. A carefully worked out code underlay each transmission—giving the partners a picture of each other's successes or failures. So far, it had withstood the test of the years. Their reputation as highliners in the trolling fleet remained undisputed.

Within the fleet a dedicated espionage team spent much time and energy trying to crack Jerry and Oscar's code. At the fish packers, a cat and mouse game accompanied each unloading of the partners' groaning hulls. As a huge catch was exchanged for a fistful of receipts, they parried penetrating inquiries with the glib edge of their tongues—giving out enough names of fishing holes where the catch had supposedly come from to fill up a good-sized atlas.

Ah yes! They were a slick pair. And the fleet never took their eyes off them.

Then Jerry's sister fell for a ne'er-do-well, marginally successful fisher by the name of Dwayne—"Dwayne the Brain Drain" as the newly suffering Jerry dubbed him. But with typical west coast hospitality, he put aside his feelings and welcomed Dwayne to the family fold. His sister's wedding was a definite sinkhole in the formerly level plateau of family relations.

When Jerry's sister approached her brother about showing Dwayne "the ropes," he knew he was in a tight place. "The ropes" could only mean how to fish, when to fish, and more importantly, where to fish. He shuddered inwardly at the thought of explaining this scenario to his pal Oscar.

Oscar won the prize for magnanimity.

"Well, he is your brother-in-law. You've got the family to make happy—we'll make out fine with one more."

This good-hearted and chivalrous remark would taste like burnt toast before the season was done.

They explained their secret code to the all-but-drooling Dwayne, explained the system which backed it up, and explained the vital need for this information to be shared only among the three. Dwayne readily agreed—and with cash registers ringing in his ears, prepared his boat for his share of the golden trio he had just been initiated into.

"If we remain friends through this season," sighed Jerry to his pal, "We'll be friends forever."

Oscar patted him solemnly on the shoulder, and they began to get the boats in readiness.

The salmon seemed to be in most of their old familiar haunts. The two partners were beginning to come to terms with the family inheritance Jerry had been saddled with. Then suddenly one morning Dwayne's brother Clarence from Bella Bella appeared, fishing alongside the three as if he knew no shame.

Heated consultation with Dwayne revealed the Unworthy had felt a rush of sibling protection. He had a duty to help out his brother, who was brand new to the game—and yes, he had told Clarence how the radio code operated. Crime turned to outright infamy as his cousin from Prince Rupert appeared, then his uncle Herman from Lund and his uncle Herman's son from Texada Island.

Oscar and Jerry ran out of patience long before Dwayne ran out of relatives. They tried splitting up, but there were enough vessels in Dwayne's navy to dog their every step. The two former highliners held a council of war. Despite the financial consequences they resolved to leave the remainder of their cherished hot spots secret, take a bath and live to fight anew next year.

As a result, the puzzled fleet of radio monitors heard not a peep from the embattled pair. Cobwebs had grown thick and luxurious on their radio handsets as the summer drew to a close. Dwayne and his family associates all fared poorly as well.

In fact, Dwayne's frustrated and cash-strapped wife made him sell the boat—and found him a job earning real money driving a cement truck. He managed to become marginally mediocre at this endeavour too. His absence from the ocean's waves was greeted by two individuals as joyously as the merchants of two centuries ago greeted the news of Captain Kidd's hanging at Execution Dock.

As the secrecy of their radio code had been rendered into tatters, Jerry and Oscar spent the dismal winter meticulously working out a scheme of visual signals. For instance, if Oscar scraped and scrubbed at a frying pan over the side of his boat, it meant, "Frig this foolishness, I'm heading over to the Brandt Banks." If Jerry hung a spotted dishrag on the mast stay, it meant, "Well, there's sweet-all happening here, I'm running for the hole on the south end of Bothman's Island." It was sophisticated and smooth—and their minds soared into the stratosphere in efforts to fine tune it.

As a fresh season began, the fleet quickly became aware that signals were being exchanged between the two boats. Collectively, they tried unsuccessfully to fathom them. Individually, they all felt they had ciphered it out. So, nervously, whenever some article was displayed on Jerry or Oscar's boat, it would be accompanied by a great-hauling-in-of-gear and scattering-to-whatever-fanciful-point they interpreted the signals to mean.

Oscar was beginning to have fun. He instructed his wife to attend some garage sales and find him some useful bric-a-brac. One day on his wire stays, he artistically displayed a paper Japanese parasol, a carnival Kewpie doll and a bouquet of brilliant red plastic flowers. After the fleet had frantically dispersed, the two friends found themselves alone for two long, productive and rewarding days.

Oscar's ever-changing flea market on the afterdeck was too much for the fleet. It kept the pair's intentions a puzzling mystery. Jerry had learned to interpret only the previously agreed-upon series of messages. However, unintentionally, Jerry ended up playing the biggest card in the deck. He washed some laundry in a bucket with saltwater soap, rinsed them over the side, and innocent of any intention other than drying the articles, hung them upon various stays and wires around the hatch. The brightly checkered shirts and coloured Stanfields resulted in the same consequences one could expect from the insertion of a stick into a hornet's nest. Boats headed in a hot fever for every point on the compass.

They were alone upon a smiling sea.

Closing up closer to his partner, Oscar gleefully discussed the great fishing, which seemed to have improved tenfold—seemingly as a reward for allowing the salmon some breathing room. He made the following observation:

"Near as I can calculate from the directions they've taken, one third of the fleet is headed for Chile, another third's halfway to Japan and the rest is northbound to Alaska. Think I'll do a wash tomorrow."

Clean clothes, full holds and solitude blessed the duo for the remainder of the summer. 🐟

Sidney

SANDY JONES AND THE *KAWASEMI*
The Last of Their Kind

Rob Morris

When you see the troller *Kawasemi* tied to the dock in Tsehum Harbour north of Sidney, BC, you are looking at a little piece of commercial fishing history. The first thing you notice are her lines. She is one of those small, beautifully proportioned double-enders that can draw your eye and cause you to pause while walking the wharves of this coast. *"Kawasemi"* means "kingfisher" in Japanese, and her slim, graceful lines certainly evoke an image of that delightful, winged fisher.

There used to be hundreds of the double-enders, each with its tiny, horseshoe-shaped wheelhouse set on a forward trunk cabin: many were originally built as gillnetters to fish Rivers Inlet and the Fraser and Skeena rivers. But, unfortunately, as representatives of this coast's fishing and wooden boatbuilding industrial heritage, they are becoming an increasingly rare sight as the years and changing fisheries and new vessel designs take their toll.

Look a little closer at *Kawasemi* and you are seeing a cod and salmon troller with a 1940s-vintage gear set-up. She's got one of the early versions of a mechanical drive for the gurdies: an automobile transmission and V-belts driven off the engine flywheel. And you do not see wooden trolling poles in the BC fleet these days, given the strength and practicality of aluminum. But there is *Kawasemi*, with a set of slender, recently painted cedar sticks.

Kawasemi on the ways at the Oak Bay Boathouse, 1960. Photo: Pete Coleman collection.

Fishing vessel licensing has its own history. What was not outwardly apparent about *Kawasemi* was the fact that she and her owner of the last twenty-five years, Sandy Jones, held the last salmon "B" licence on the coast.

SANDY JONES AND THE *KAWASEMI*

Kawasemi at the Tsehum Harbour float at Sidney in 1996. Photo: Rob Morris.

Part of an initiative to reduce the salmon fleet in the late sixties was to divide the general licence into three commercial categories—A, B, and C. These were issued according to the fisherman's catch record. An A licence was issued to fishermen with a catch record of over 10,000 pounds of salmon per season; it also allowed its owners to fish a variety of other species. This licence was hard to maintain, as it required a mimimum seasonal poundage in order to be renewed each year. A B licence was issued to fishermen with salmon catch records below 10,000 pounds, and had a life span of fifteen years from the date that it was issued or traded. It was intended to ease fishermen who did not concentrate primarily on salmon out of the salmon fishery. Most of these "interim" B licences were meant to roll over to C licences when they expired. C licences excluded the right to fish salmon, but allowed the retention of certain other species; they were issued to fishermen with no history of salmon landings. In any case, there were 643 Bs by 1975, only eight by 1985, and, by 1989, as Sandy Jones was able to say, "Now I'm the only one left!"

Kawasemi does not appear in the Ships Registries, so her origins are not documented. Sandy recollects that she was one of three sister ships built in 1940 for gillnetting up in Rivers Inlet. Tor Miller, a veteran Victoria lingcod troller, remembers *Kawasemi* when she belonged to Cecil Joyce, his fishing partner for many years in the Strait of Georgia. "She came out of the Japanese fleet that was impounded in 1942, and Cecil took her dog salmon

One of Sandy Jones' fine double-ended rowboats. Bruce Burly of Victoria is at the oars in this photo taken in the early fifties. Photo: Bill Grundison.

This view of the *Kawasemi*'s deck shows the Dodge transmission beside the hatch and the belt-driven trolling gurdies. Up against the cabin is the belt-driven cod gurdy. Photo: Rob Morris.

fishing in the Queen Charlottes, using those sunken gillnets. Then he took her cod trolling in the Strait." The *Kawasemi* still has the live-well set-up from those days, with pluggable holes through the hull and gurdies amidships, with a starboard davit for trolling lingcod. Tor Miller recalls Cecil Joyce saying that *Kawasemi* was "built in the Cove by Atagi." Cecil hailed from Quathiaski Cove on Quadra Island. Ken and Bill Atagi, owners of Atagi Shipyard in Richmond, recall their uncle Kuomatu Atagi shipwrighting at BC Packers in Quathiaski Cove around 1940, but don't know if he was the builder of *Kawasemi*. When his partner died, Tor sold the *Kawasemi* to Peter Coleman, who fished lingcod and salmon with her in Haro Strait from 1958 to 1964. Pete sees her at the Tsehum Harbour wharf on the way down to his troller. "She's still got that same green and white paint job as back when I had her," he laughs. *Kawasemi* then went to Art Albany of the Songhees Indian Reserve near Victoria, then to Sandy Jones.

Kawasemi and Sandy trolled lingcod, as he had started doing with his dad when he was thirteen years old, and salmon at Active and Porlier passes, around Pender Island and out of French Creek. "I used to buy a licence for a dollar and fish year-round," he says. "I did good on winter springs."

The first engine in *Kawasemi* was a 10-14 Easthope. "I used it for a couple of years but it was too slow, so I put in a three-cylinder 15-18. Oh, it used to go! I had it for six or seven years and then I put this high-speed in. I sold that Easthope for $500. I'm kicking myself; I let it go! I used to troll all day on 5 gallons of gas; this one takes over 20. So I play it safe with that extra 65-gallon tank on deck. It's a Chrysler Ace 85 horsepower."

Sandy explains the drive system for his gurdies. From the cockpit a foot pedal is connected by trolling wire to a belt tightener around the engine flywheel. This engages a shaft running aft, along the deck to an automobile transmission which turns a long V-belt engaging both sides of the gurdies. "I'm the only one that's got

that old style; everybody is hydraulic," Sandy laughs. He is more comfortable with his Dodge transmission. "I just use low gear. In second it all comes up too fast."

Leaning back to look up to the tops of his trolling poles, Sandy comments that they are only a couple of years old. "Cedar," he says. "Same length as the boat (32 feet). Should be good for five or six years. With fir poles you can go ten or fifteen years." He loves and wants to maintain the original lines of *Kawasemi*. "I'm going to rebuild the wheelhouse. I've got another one which would go on, but it's higher and would spoil the looks. She's a good sea boat. She's cedar-planked and I've never corked her since I've owned her. Every year I check her on the ways at Canoe Cove."

A few years back, Sandy got sick and sold his A licence. Then he tried fishing lingcod again and found he was having trouble making a living, so he appealed to the federal Department of Fisheries and Oceans. The Fisheries minister granted him a lifetime B licence which would expire and go to a C when he retired from fishing. As of 1996, he was still the only B licence holder on the coast.

Asked when he thinks he might retire, Sandy just laughs. "I like fishing! I got out last summer, but I don't fish like I used to. I'm over eighty years old, you know—get tired pretty quick!"

Sandy Jones. Photo: Rob Morris.

Illustration by Graham Wragg

LOVEBITES

a short story by Jean Rysstad

Paul and his father are coming in from a ten-day trolling trip. It has been a long ten days without much fish. Five or six in a seventeen-hour day. These northern summer days stretch from five in the morning until near midnight. "Scratching" is what they call this kind of trip once they are in town. That's how Paul's dad will sum it up on the dock to anyone who asks. Paul hasn't had to use the term "scratching" yet this season because until now they've done well. He'd like to be able to be as casual about a bad trip as his dad is, but he thinks he'd feel pretty phony, that it would sound fake coming from his mouth—as if he can accept this kind of a trip, that it rolls off him like water.

For Paul, it has been one of the worst trips he can remember. He longs to be home just to have a bath, use the telephone, lie on his own bed instead of a bunk.

The thing is, Paul's dad wants the fish, no matter how many or how few, handled with care. He wants them babied. His dad wants the fish to look and feel like they are still alive, and it requires a dedication that Paul doesn't feel this year although he wants the money from the fish more than he has ever wanted it, and he wants *more* money than he has ever wanted. Though Paul is satisfied that he's done his job, gone through the motions, sharpened and baited, cleaned and iced the same way he always has, he's disappointed both in the trip and in himself.

He has always thought he had what it took to fish, but now, halfway home, coming across from the Charlottes, he feels like he'll go crazy if something doesn't happen to make the trip go faster. He'd like to be able to say, well, at least this or this happened. But he doesn't know what it could possibly be. The wind is light and the sea is calm. The weather voice promises more of the same.

Paul shrugs. He guesses he can make it home in one piece, body and soul intact. He guesses he'll live. He'll make it through not just this trip but the next and the next and the next until the season's over. It gets harder and harder to have a decent trip—or so he's heard until he can't stand it any more.

His dad could retire but he keeps on fishing. Paul thinks that his dad expects or hopes that Paul will take the boat when he's through school, and Paul thinks or hopes that too. But for the last few days, when it was obvious they weren't going to get much, Paul's been thinking his dad would be better off alone than with him around. He wouldn't have to pay out a share to Paul.

At the galley table below deck, Paul considers the V-shaped birds in flight design he's drawn on the note pad they keep handy there for grub lists, for marking down gear or lures they want to pick up in town. He's been using a pencil on a string tied through the hole in the shelf above the table. He drilled the hole one trip when the only pencil he could ever find rested on his dad's ear.

On this trip not even the weather has been anything to speak of. Usually there is a day or two where they have to battle with the weather, and Paul doesn't mind this. Then at night, dropping anchor, holing up, is a small celebration. Paul makes hot chocolate. He reads. The bells jangle on the stabilizer poles. The wind is a low-pitched humming song on the rigging.

Paul reads *Popular Mechanics*, *Consumer's Market* and *Omni*. He likes to compare brand names, performance data of the various items he considers buying with his crew share. He's spent much of his time this summer daydreaming about what he'll be capable of buying. He'll be able to get his driver's licence soon, and a car is number one on his want list. On the last trip, Paul had been comparing Hyundais, Toyotas, small Fords and Chevs. Opening and closing the two-versus-four-door vehicles in his imagination. Definitely two-door, he'd decided. Classier. One for himself and one for a girl whose name he wasn't sure of. It wouldn't be Trish or Kim. He wasn't comfortable with girls who had names like that, and most of them did. Cathy or Janey was better. He had a composite picture of this girl drawn from the fifteen or so girls in his grade ten class—taking this item of clothing from one and this way of standing from another. It was a kind of composure of body and mind that he dreamt about in a girl. Or in himself, Paul thinks. He swats the pencil on the string so that it swings wildly back and forth.

Paul goes up to the cabin. His father, at the wheel, greets him, lifts his cap with one hand, and runs his fingers through his hair with the other, then sets his cap back on. Paul catches himself about to repeat the same gesture but forces himself to stop midway. He doesn't want to be exactly like his father, a copy of his dad, with the same kind of calm acceptance of everything, going steady in one direction until it's time to go in another. Paul wishes his dad would do something erratic, surprise him, shock him, but he never does and never will. Without thinking, Paul completes the gesture he consciously stopped himself from making. He realizes what he's done only when he's readjusting the peak of his hat to sit firm and tight on his head again.

"Shit," Paul mumbles. It is hopeless for him to try to get back into the dream state for the travelling-in. The dream girl, who last trip wore a blue flowered skirt, those ones with a shy bit of ruffle at the hem, had vanished. The possibility of her was gone. Oh yeah, he remembers with scorn, she was going to be the kind of girl who found everything interesting. Who never used the word "boring" or "bored." She was the kind of girl who'd like trolling, who could laugh at herself and amuse herself and not need to be talking all the time.

It was only five days ago that Paul had decided that he should check the daily newspapers, the buy-and-sells and the used car lots to see what was around for cheap. Say five hundred to a thousand dollars. He considered this a mature, realistic assessment of his financial situation. He figured he and his friend Trent could work on this car in mechanics shop at school and at home in the garage.

"I'll take a wheel turn," Paul offers. His dad doesn't hear him. Paul stands closer. "Do you want me to take it for a while?" Paul asks again.

"No," his father says, "but you could put the coffee on."

Paul doesn't like drinking boat coffee or making it. Cold water and a handful of grounds in a saucepan. They use the propane two-burner for coffee, not the oil stove below. Paul watches that the brew doesn't boil for more than a second, then pulls the pan off, adds a touch of cold water, and watches the grounds settle.

"About three," his father says, meaning his estimate of when they'll get into town: 3:00 p.m.

Three more hours is what it means to Paul.

"Do you want something to eat?" Paul asks, hoping his dad will say yes. It will give him something to do.

"A couple of biscuits would be fine," Paul's father says. "That'll do. We'll let your mother know we're coming in a bit."

The Digestive cookies Paul's father likes are in the cupboard above the burner. Paul passes the box to his dad, who fumbles in the corrugated envelope and comes out with the last three. "That's the end of that package," he says with a grin. "Want one?" Paul hates Digestives but he'd like one now. His father knows how to ration, but Paul's eaten all his treats early this trip.

Paul twists the rectangular bag into a figure eight and shoves it into the garbage can. "You're sure you don't want me to do a stretch?" he asks. His father shakes his head and slurps a mouthful of coffee. It seems to Paul that his father knows how much he'd like to take the boat. His father tests him in little ways all the time.

"Okay," Paul shrugs. "I guess I'll try and dig up something to eat."

All the envelopes of hot chocolate are gone now. There is a can of strawberry milkshake powder in the back corner of the cupboard but Paul gags when he looks at it. The milk, on ice in the hold, is past date. Paul hates that chalky past date taste.

His mother will likely put a roast in the oven when she gets their call, and he wants pizza. Or maybe Chinese. He closes his eyes and sees the take-out menu. Combos One, Two, and Three. NO SUBSTITUTIONS in bold print. He knows he shouldn't test his sanity like this. He's going crazy, climbing walls. Three more hours of himself.

He sits in the cabin doorway, where his back gets the wind. Facing his father's back, he begins a motion. A pushing and rocking forward, pretending his pushing will help them move faster. His father seems to feel this urging behind him because he glances over his shoulder at Paul. "Won't be long," he says. "We're making good time."

Paul stands. He doesn't know why he wants to be in town so badly. It's not like he'll have any time to himself. They have a lot to do and they will be back out within two days if the weather co-operates.

He takes a leak over the side of the boat, surveying first the sky and then the arc of his water. It used to be more fun.

He zips and turns and looks at the cabin, the roof, and he remembers when he used to sit up there on nice days, feeling like the luckiest kid in the world. What the hell, he'll climb up there now.

He sits yoga style then lies down, watches the grey sky and the grey water from this vantage point. He remembers how high it used to seem, how dangerous, when he sat up there as a kid. How awed he used to be by the depth and width of the ocean. In bad weather he used to be afraid, but his father never tried to help him over or through the fear. Or maybe he did, by just letting him work it through for himself.

Paul hears his father call his name. He could reach down and knock on the bow window but he doesn't. He doesn't answer the next, sharper call either.

Paul is not sure why he isn't answering immediately, and one part of him, the part that is in this world, not lost, overboard, cannot believe his own silence, his refusal to answer his father's call. He cannot believe himself when he doesn't scramble down and say, "Hey Dad, I'm here," when he sees his father stand helpless for a moment at the stern, then hurry to the bow, searching off the side of the boat as he runs.

The boat swings around and Paul has to grab the lines that hold the small dinghy on top of the cabin to keep himself from falling off. He does not think how ironic it would be if he fell over then, when his father turned the boat. Later, it occurs to him.

He climbs down quietly. Stealthily. He knows it is too late to reverse his disappearance though he hasn't been missing for more than ten minutes. Maybe less. His father would be thinking ten minutes was both a long and short time too. Would be scanning port and starboard with binoculars, first with his glasses on, then off.

Paul speaks loudly from just inside the cabin. "Dad," he says, "it's okay." He walks towards his father because that's where his feet take him. When he comes within arm's length of his dad, he shrugs. He doesn't want his body to shrug. It shocks him.

Into his father's eyes comes first a glossy black welcome like hot tar, thick as blood, which Paul wants to return which Paul wants to reach for. If only eyes were arms.

But his father's eyes empty at the shrug. The colour cools, flattens.

His father takes the boat out of gear. He grabs Paul's shoulders, bony under the red and black flannel workshirt and faded black T-shirt he wears, and shakes.

He shakes Paul by the shoulders and speaks. Staccato:

"If. You. Ever." Jolting the boy with every word. "Do. That. Again." Bones, teeth, brains. "I'll. Kill. You."

Paul's father's hands drop and he turns away from Paul, lifts his cap, puts the boat in gear, ten-knot running speed. He takes a mouthful of the cold, grainy coffee Paul's made for him and spits it out the window.

Paul backs away, trying his shoulders. First one, then the other, in the small warm-up circles he uses before a workout or a run. He lies below deck on the bunk, eyes closed, ears tuned to the drone of the engine. He falls asleep.

When they nuzzle up to the plant dock—as always, the perfect gentle landing—Paul wakes. He thinks he might tell his father that he'll sleep on the boat. That he'll unload, clean up the boat, get ice. He could ask his dad to let him. That way, his dad could have a rest, forget about the boat for a day or so, and he could try and make it up. But he doesn't have the nerve to say anything. He does his usual share. Hoses down the deck and scrubs the checkers, gathers the dirty clothes and the garbage. His mother's waiting above the dock in the pickup. He steps off, knowing he'll have to take his place, seated between them. He walks like he is still on the boat, his sense of balance in transition.

After his father and mother go to bed, Paul runs the bath water. He brings the small television in from the kitchen and sets it on the shelf above the taps, pushing all the varieties of shampoos and rinses and little pink and blue seashell soaps to one side. He hopes the TV won't fall in the tub. Then again, it might be good if he got zapped, he thinks. The water pours down hot and clean on his big stinking feet. He dumps a half bottle of his mother's bath foam under the waterfalls, scoops the foam and blows it. Turns the channel around once and stays with the late night show host, Letterman. The band leader named Paul is the butt of Letterman's jokes, which is fine with Paul in the tub. He deserves it. There's an Irish actress whose talk is off the wall. The things she says seem to come out of nowhere, from nothing said before or nothing Letterman's asked her. Paul scrubs himself everywhere and gets out of the tub. He examines his back and shoulders in the three-way hinged mirror above the sink. Eight purple-black-and-blue imprints blossoming like buds. Lovebites. Paul snorts. A subject for Letterman if ever there was one.

He rubs himself with the towel, which he had hoped would be soft, fluffy, but it isn't. Dried on the line, as usual. His mother still thinks towels and sheets dried outside smell so fresh.

Paul dries every toe, every crack of himself. He's begun to feel better, but what is Letterman doing now? Crushing a tube of toothpaste under a ton of weight. The paste spurts out from all the seams. And now a doll. A child volunteers her doll for the experiment, and she stands puzzled, frowning as deeply as an old woman.

There is a knock at the bathroom door. Gentle. Paul snaps the TV off.

"What?" he says. It is his mother's knock. "What?"

Her slippers shuffle. "I just have to tell you this."

There is a pause between them. "Once, when I was old enough to know better, I pulled a chair out from beneath my grandmother."

"And what?"

"She fell hard, Paul. She broke her tailbone."

Silence. "Paul?" she asks.

"Okay," Paul says. "Okay, Goodnight. Goodnight, Mom." 🐟

Excerpted from *Travelling In* by Jean Rysstad, 1990. Reproduced with the permission of Oolichan Books. Published in the *Westcoast Fisherman*, 1988.

Ucluelet

FISHING IS A QUALITY OF LIFE
Ucluelet Trollers Ralph and Judy Hodgson

Valma Brenton-Davie

It is summer and the height of salmon season. In Ucluelet, on Vancouver Island's west coast, Ralph and Judy Hodgson prepare for another ten-day fishing trip.

The couple own the 41-foot *Tonga*—a classic built by the famous Wahl yard of Prince Rupert—and have fished together for twenty years; about as long as they've known each other.

Judy is heading up to the co-op store. She wheels half a dozen fold-up totes. "The co-op here never has any boxes," she shrugs. And Judy should know. She and Ralph have fished out of Ucluelet for eighteen years. As she shops, she talks about how she and Ralph entered the commercial fishing business.

"Ralph had been running fishing charters for seven years. Me being a prairie girl, I wasn't quite so familiar with the sea." That was soon to change. "On one of our dates," she reminisces, "Ralph came to pick me up with his boat at the Planetarium in Vancouver. I was sitting beside the crab sculpture. When Ralph arrived, he said, 'You remind me of a mermaid. If I buy a fish boat will you go fishing with me?'"

To this, Judy answered, "What will you pay?"

Ralph and Judy Hodgson. "Jude and I are just a Ma and Pa operation," says Ralph. "I believe in the old-time fishin'—bare necessities of equipment. We just go out and do the best we can." Photo: Valma Brenton-Davie.

TONGA
VANCOUVER BC.

FISHING IS A QUALITY OF LIFE

The *Tonga*, alongside in Ucluelet. Photo: Vulinu Brenton-Duvie.

Ralph answered, "Depends on what we make."

Ralph and Judy walked the wharfs during the fall of 1974 and finally purchased their first fishing boat, the 36-foot double-ender *Ocean Prince*. The couple started their new life together day-trolling in Area 6 on the mid-coast. They fished the *Ocean Prince* for four years and settled on Ucluelet as their salmon-season headquarters. At that time, Judy remembers, there was a day-fleet of about 150 boats and it was quite a sight to see them all heading out of the harbour at once.

"We didn't have much equipment on the *Ocean Prince* and it was my first time steering a boat," Judy explains. "People would ask Ralph what kind of autopilot he had on the boat. Ralph would reply, 'A Judy!'"

A few years later, the Hodgsons decided that they needed a bigger boat, one that could pack ice so they could make longer trips. In the fall of 1978, they purchased the *Tonga*. "Ralph started overhauling the stern and worked his way to the bow, stripping everything down," says Judy. By the following spring, the *Tonga* was ready to go to work again.

Since then, Ralph and Judy have fished the *Tonga* together, and only occasionally over the years have they taken a deckhand. "Ralph is independent," explains Judy. "He marks his own lines, ties his gear and pulls the gear. In fact, there was a lot of years that Ralph did everything in the cockpit, even eating his meals."

She smiles, "We have a pretty good system on the boat. I look after the pointy end and Ralph looks after the square end, the middle is common ground." Judy only goes to the stern to throw the gear during sockeye.

Back at the boat, Ralph sits in the wheelhouse and listens to the VHF. Outside, southeast winds are increasing and a light rain is falling. The radio predicts southeast gales reaching 50 to 60 knots at Cape Cook. "I'm not heading out in that crap," says Ralph. "We'll be tied up here until she settles down."

Ralph lands a salmon while fishing La Perouse Bank off Ucluelet. Photo: Hodgson collection.

With the groceries stowed, Ralph and Judy relax in the *Tonga's* warm, spacious galley. Tea is made and with the boat swaying gently, Ralph explains that to him, more than anything, fishing is a quality of life. He has always fished in one form or another to make a living.

At nine years of age in Prince Rupert, he acquired his own leaky rowboat which he caulked with rags. "I loved to tie my own flies and I would take kids and adults out fishing in my little boat, charging them ten cents per trout. Six trout, that was 60 cents—a lot of money back then." The money eventually went to buying a charter boat, which Ralph operated for seven years. It was then that he met Judy.

Although Ralph is a skilled pipefitter, he decided that commercial fishing would offer the quality of life that he wanted for himself and his lady.

"Jude and I are just a Ma and Pa operation," says Ralph. "I believe in the old-time fishin'—bare necessities of equipment. We just go out and do the best we can. We don't like to get caught up in the stress and politics of the way fishin' has become."

Like many fishermen, Ralph and Judy are concerned about low prices for their fish. The Hodgsons feel frustrated that they have invested twenty years in the business and are now wondering what will happen to the quality of their chosen lifestyle. At the same time, they remain optimistic. They both have other skills and know that a day may come for them to make some changes.

For Judy, twenty years has gone fast and she says that if she had her choice, "I'd do it all over again."

The Hodgsons relaxing in their comfortable galley. "We have a pretty good system on the boat. I look after the pointy end and Ralph looks after the square end, the middle is common ground." Photo: Valma Brenton-Davie.

SO YA WANNA BE A HIGHLINER?

Billy Proctor, ex-highliner

This is what it's like to try and be a highliner. You have to:

- be the first boat on the grounds in the morning and the last one to leave at night;
- find the fish before the other guy does;
- when you do find them, they have to bite better than the other guy's;
- when the bite's on, you have to work faster than the other guy;
- don't stop to eat as often as the other guy; better yet, don't eat all day.

Most of the time, the above rules will give the little edge you need to beat the hell out of most guys. If you need more ammo, try the following:

- in the interest of finding more fish than the other guy, try to find out what kind of lure he's using without telling him what lure you're using. If you do tell him, learn to lie;
- lying is a necessary phone skill: "Got a few yesterday, but not much today," or, "No good here, too many jellyfish";

THERE! NOW YOU CAN GO TO THE PUB!

Illustration by Elly Loven

- when the bite is on and another boat comes by, go sit in the wheelhouse and hope he doesn't see the poles shaking. Or, if he stops to holler at you, tell him,

> a) you haven't had a bite in hours,
> b) you're just going to pick up and move,
> c) you've heard there are lots of fish over at ...;

- when you find a killer spoon, plug or hoochie, guard it with your life: I knew guys who would take them to bed with them;
- when you come in with a good trip, change all your gear and hide all the good stuff and put any old thing in the stern;
- if anyone asks you where you were fishing, tell him someplace miles from where you were;
- make sure your deckhand doesn't go talking to another highliner;
- share your "killer" plugs with fellow highliners (the plugs that don't catch a thing), ask if he has anything like it. Good strategy for seeing his hoochie box, if he shows it and not the "junk box";

- take a real killer hoochie to the gear store and buy out all they have;
- if you come in with a really good trip and there'll be guys watching you unload, get your ice, grub, and fuel first;
- tie up somewhere easy to get out. Wait till the other guys leave for a beer at the pub, then pull out and anchor somewhere;
- turn your phones off when you're into good fishing; then, if someone says, "I called you and got no answer," you can say (honestly) that you never heard him (this can backfire if he's into better fishing than you are);
- disregard the BS factor; someone will always say they're catching more smileys than you are;
- keep a good number of excuses on hand when it turns out you're not the high boat for the trip: "Wrong place," "Speed was wrong," "Water was too clear/cold," "Too many boats," "Got tangled up when the bite was on/lost two hours of the best fishing," "Bad roll of perlon/box of hooks." BS excuses, of course, but they make you feel better;
- when all other excuses fail, use the catch-all, that there's something wrong with the boat: "Don't know what it is, boat fished good yesterday, won't catch a thing today."
- deckhands are also very popular to blame your troubles on;
- shadow the guy who does better than you when he goes to the gear store. With any luck, you'll see the killer gear he buys;
- look for near-empty hoochie boxes in the gear store. Must be some good reason they're near-empty (doesn't always work out, but often enough).

You-know-what aside, there are some absolute MUSTS to remember if you wanna be a highliner:

- keep your hooks sharp;
- keep them clean at all times (can't catch fish if you have jelly-fish or junk fish hanging on your hooks);
- keep your leaders straight at all times;
- try to get more tacks along the hot spot than the other guy (but make sure you don't cut him off or get in his way);
- always be fair, no matter what. 🐟

Illustration by Elly Loven

'CLICK' 'CLICK' 'CLICK'

THIS IS MY KILLER PLUG!

JOAN LEMMERS
A Woman's Life at Sea

as told to Vickie Jensen

My Early Years

My dad started commercial fishing when I was six years old. He went away the first summer with my uncle's old boat, a little bit of fishing gear, and something like $28 for fuel and food. He'd never commercial-fished before and a wonderful native man took him under his wing. My father came home with a pillow case full of money at the end of the summer. We sat on the living room floor putting it all in little piles to see how much money he had earned. But it was all of the adventure stories that he told that probably really got us hooked. That and going on the boat, riding up as far as Campbell River with him.

From then on my younger brother and I always went out on the boat during the fall and winter months fishing out of Horseshoe Bay. My mom got seasick all the time so she didn't like going along. We shovelled snow off the deck of the boat and used it to ice down the fish. And helping paint the boat in the spring was always really neat.

When I did go along, I wasn't allowed to run the gear because it wasn't girl stuff. I could choose which was going to be my line and I could choose what gear I wanted to put on it, but my younger brother had to run it or my dad had to tie all the gear up.

In the wheelhouse, Dad would show me how to use the radar and how to navigate, how to watch the depth sounder, but not the gear. I might get hurt. I wasn't strong enough. Or I wasn't tall enough. The fish would be too big. It was okay to put me on a wheel watch for two hours, but I couldn't run the gear.

The bad part came when my brother, who was only twelve, and my younger cousins got to go fishing with my dad. I wasn't allowed because I was a girl. I was fifteen so ended up working in a cannery down at the foot of Carrall Street. In a way, I came out ahead of the game because I was on union wages and getting the same wage as a man with lots of overtime. I had a horrible job at a horrible cannery, but I was a rich kid.

My Own Boat and Family

Eventually I grew up, got married and had my first baby. By the time my son was two, Dad taught him how to use

After she bought her own boat, Joan Lemmers finally got to run the gear. Photo: *Westcoast Fisherman* collection.

everything on a boat. He could start the boat, talk on the radiophone, turn on the radar. Then Dad sold his boat on the first buyback program and retired way too young. He went crazy not having the boat.

I was also going crazy without a boat because there was always this wonderful social life down at False Creek. All of a sudden there was no boat to paint, no reason to walk around on the docks and talk to people.

My dad and I looked at a terrible old boat for sale, an old gillnetter. You could see right through the planks. But Dad assured me, "Ah, you just mix baking soda and cement and just patch it all up." For $800, I got the boat, a net and a B licence. The engine started and off we went. I had never gillnetted before.

That boat didn't even have a water tank if you wanted a drink, and the only dry spot was the driver's seat. You slept in your rain gear. It was so unreliable and we were so poor that every opening we'd take the Willard battery out of our car and carry it down to the boat as a backup. So it became a family joke that before we untied the boat, someone would always ask, "Is Mr. Willard with us?"

In the beginning, we'd just fish off the mouth of the Fraser. Then we got really adventurous and went over to Texada Island with the two kids, Michael and Suzan, when they were about four and seven. My husband Rob taught industrial education full time at West Vancouver Secondary, so both the kids fished with me. They loved going out on that funny old gillnetter. It was a big adventure to them. They wanted to go every single time. You'd think it would be a bit boring for them because it's mainly fishing at night and there isn't that rush bringing in the big salmon like when you're trolling.

More Boats

Eventually Rob and I decided to build a new boat, a troller. Now that's another story. We would work on it until our money ran out, then put in another fishing season with the old gillnetter and get enough money to buy some more wood or a used engine. I think it took us four years to finish the new boat. I would never, ever be part of a building project again unless I was at least a block or two away from the scream of the power planer and the clouds of sawdust!

Our daughter Suzan was probably around twelve when she fished with Rob and me on our new boat, the *'Ksan*. She became our full-time deckhand when she was about fifteen. Age fifteen was terrible. Finally we just told her, "If you don't like the job, we'll take you in and you can go home because there are ten guys on the dock waiting for your job." That sunk in.

I never really saw myself as a role model until years later when my daughter came along and said that she wanted to buy a boat. I said, "Where did you ever get this idea?" She told me that if I could do it then she knew that she could do it. I hadn't thought about myself as a role model at all. I was just busy doing it.

Even so, I advised her to just forget about buying a boat. She said,

"All I want you to do is co-sign the loan." I argued, "No. Just forget it. You've got a good job." So she called me a sexist pig because I signed for her brother's boat and I wasn't signing for her boat. She told me her boyfriend was going to put up half the money and he had some experience on a freezer boat. She also said he was going to take a diesel mechanics course. That kind of stuff. So I told them to find the boat, do everything, and then tell me about it and I'd co-sign. In the end they got a really dandy little boat and ended up putting in a couple of really good seasons on it.

Eventually, they both quit their jobs to go fishing. I said, "Well, what about all that university education?" They told me maybe they'd only do it for a few years and move on. Now they want a bigger boat. That's the "moving on" part.

Eventually my husband Rob quit teaching and started fishing, too. It was absolutely terrible the first year because I had been the boss and had the boat organized my way. Everything had its place. Rob is the kind of guy who will wrap a screwdriver up in an old greasy cloth and shove it in any drawer. It got to a point when we finally had to say, "Okay, there can't be two skippers on the boat. Who's going to be the boss?" So it was decided at that time that I would be the skipper. But in the last few years, I've sort of stepped back and Rob's taken over.

Rob and I have noticed that from age thirty to forty or forty-five fishing is really easy. But from forty to fifty we've both found our energy level isn't the same. I can still really push it, like doing a wheel watch all night, because I don't need much sleep. But last summer there were just three of us on a freezer troller during sockeye, so we had to work flat out for several weeks. You go to bed and your hands are all swollen up. You wake up and you're just as tired as when you went to bed. Your hands are still aching and all puffed up. You can hardly get them moving. And you think, "Do we want to do this forever?"

Sometimes I think maybe I won't fish full-time. If it's pouring with rain and it's howling and there are no fish and I'm missing all the fresh peaches and strawberries of summer, then I think, "God, there's got to be an easier way to make a living. But other days it's nice weather, the sun is out, and you're off the west coast of the Charlottes with whales and dolphins and eagles, and the fish are biting and you think, "Wow. This is it. I don't want to be anywhere else. This is what I want to do forever."

Safety and Close Calls

I've got a thing about safety and safety equipment. For example, if crew members use a flashlight, they've got to put it back right away and they've got to put it in that place where it's kept because I have to know it's right there at all times. Once I had an engine room fire. I realized the first thing I had to do was run down and throw the main switch from the batteries because it was an electrical fire. But in order to do that, I had to have a flashlight immediately so I could see to get the fire extinguishers off the wall. If I had to spend thirty seconds searching around for a flashlight, it might make a big difference. I even sleep with a flashlight under my pillow so that if something happens in the night, I can see what's going on immediately.

Twelve years ago or so I had one really bad accident when I was fishing the boat myself. I had a friend working with me, an English fellow named Graham. I guess it was only his first or second trip out. There were four of us boats all travelling together in a line, half a mile apart, all coming home to Vancouver. I was number four, last in line. I'd been at the wheel six hours peering through this thick, heavy fog and was just exhausted. You couldn't even see the bow. We had reduced our speed to about half.

Finally, I had to put Graham on the wheel. I just needed to make myself a cup of tea, have a stretch and get some air. He was really scared, but I showed him what to do and told him, "Just follow those little dots in front of you and this beam. Watch the numbers and remember these boats in front of you. They're in radio contact. If they slow down, they'll tell us. Then we'll slow down, too." There was also a boat coming towards us that I'd been watching for six miles. I pointed it out on the screen and told him, "Just watch this little dot that's coming down the side. Keep an eye on him."

So he said, "Yeah fine."

About half an hour later I came back into the wheelhouse and said, "How are you doing?" He panicked and told me, "I just remembered that boat coming towards us! I forgot to watch for it!"

Together Joan and Rob Lemmers have built two houses and one 45-foot troller. Photo: Lemmers collection.

I shot questions at him. "Well, where do you think it is? Where did it go on the screen? You didn't track it or anything?" He goes, "No. No. Oh God!"

So I told him, "Okay, move out of there and I'll try and find it. Did you touch anything? Did you start tuning anything?" "No, I just forgot to watch it."

I'd just stuck my head in the radar to study the screen when all of a sudden, Crash!

For a few split seconds, we didn't know what had happened, whether we'd hit a reef or what. But we both thought it was the end. We thought we were going to die.

What happened was our boat and the other boat hit just about head on. Then our rigging got caught up in his rigging. It was an old steel tug that had been turned into a packer or something, and he was going full speed in the fog. I was at the wheel and ended up flying through the air and hitting the back wall. Graham was standing, but the impact threw him onto the stove. The whole boat was over on her side and being dragged along. Water was coming in the back door. Everything was snapping and cracking and breaking. Lines just whipping around. Graham and I just looked at each other like, "It's really great knowing you." We didn't know how we could even get out the door and through all the rigging which was in the water. From where I was on the floor, I could look out and see the back deck foaming with water and a tangled mass of broken wood and wire. It was unbelievable.

Somehow or other we broke free from the other guy. That boat just disappeared into the fog.

Fortunately our boat righted itself. But then the stove caught fire because the chimney had been pulled off. So the first thing we had to do was fight a fire. We were covered in black soot, bleeding from cuts, and trying to put out this fire.

Once we got the fire extinguished, then we had to think about sinking. We didn't know if the bow was split wide open or what. Graham was brilliant. He grabbed a fishing knife, went out on deck, cut all the scotchmen off and started lashing them together to make a raft because our canister and our rowboat, everything, was smashed or gone. The trolling poles and the mast were in pieces. You couldn't get at anything. We couldn't even radio for help.

Finally, the boat that hit us turned around and came back. The guy said, "It took me a while to slow down." Bullshit. I think they had a little discussion. First they were going to just keep going. Then they realized that maybe we were sinking. So they came back after about twenty minutes or so.

He phoned up my travelling buddies who all went into a state of shock. "They're coming back for you," he told me, "so I'm out of here."

At first the fault for the accident was put at fifty-fifty. But the other boat had actually changed direction and was going into a port, so he was crossing the four of us. And the other three boats said he did change direction in the fog going full speed. He should have known I was there and gone on the outside or behind.

I phoned Rob in tears, but he was teaching so I had to deal with everything. I contacted the insurance people, talked with all the brokers, dealt with the shipyard. The other company paid for the $30,000 worth of damage.

I think Graham and I are soul mates for life. I was always pretty safety conscious beforehand, but I think that accident made me even more so.

Woman at the Wheel

Over the years, some people thought it was strange that a woman was running her own boat, but there was also tremendous support. I had one guy come by so close that he just about ripped my [trolling] pole off. I refused to change my tack because I'd end up losing all my gear on the reef. I could see him on his big boat with his binoculars looking right into my checkers. I thought he was trying to see how many fish I was getting or maybe what gear I was using. Finally I went on the air and said, "Look. The ocean's a really big place. Why do you have to come so close to me? If you are trying to see what fish I've got, I'll tell you what my score is for the day."

He said, "Heck. I'm not looking at the fish. I just wanted to see if you look like a lady truck driver. But you're just a little shit." I said, "Yeah. I'm not that big." He was nice to me from then on. I guess he was just curious.

Then this other real rough bush-man from the Queen Charlottes came up to me in the bar and said, "Why are you doing this anyway? Obviously you don't like fishing." I asked him, "Why do you say that?" He answered, "Well, because you've got on a lace blouse."

I was wearing this heavy cotton blouse, like a man's white shirt but with lace on the front, with jeans and cowboy boots. He told me, "Fancy

"...a tangled mess of broken wood and wire. It was unbelievable." Photo: Lemmers collection.

ladies don't like fishing. They don't like the rough weather or the blood and the slime and the smell of fish."

I told him, "Well, I do happen to like fishing. I like the coast, and I like the lifestyle, and I like all the friends. I'm here because I want to be here." He just couldn't figure it out at all.

Most of the guys were great. I remember one evening when I was up at the little cabin store at Langara where there was some fishing gear for sale. This Japanese man was chatting with me while his buddies were over having a coffee. I realized he was tapping this one box at the counter where the plugs were. Finally I figured it out and bought the whole two boxes. When I came down to my boat, there was a note on my wheelhouse that said, "Very good mid-August" and the place to use them.

Or the brothers on the three *Triple M* boats who were just absolutely wonderful. One of them would phone me up and say, "Are you anchored up yet?" I'd say, "No, I'm just running in," and he'd tell me, "Well, there's a really good bottom here. It'll save you a few minutes running around in circles trying to find someplace."

Like many fishermen, I fished with a group of boats that were in constant radio contact. It's smart to do that for safety reasons, but it also makes for efficient fishing because we share information on how many fish we've got, the depth they're hitting, effective lures, water temperature, and even recipes! It was like having a dozen big brothers all taking care of me. Every single night I had to phone in and say where I was and that I was anchored up safely and everything was okay. They wanted to know was the boat running properly and was there anything I was having problems with? It was really great.

Fishing and Politics

I didn't get involved in the politics of fishing until about fifteen years ago. I was one of the few women regularly attending the Pacific Trollers Association [PTA] meetings and one day I had a phone call from a woman working in Ottawa. She was furious because she'd seen a letter in which the association told Ottawa there were no women within the industry who were capable of serving on the Salmon Commission.

I went into our next branch meeting, waving a copy of that letter and shouting, "You jerks! The world is changing! There are women who run fishing companies. There are women running packers. There's a woman who has her own troller and she's a lawyer. You're telling me that a lawyer can't do a job on a salmon commission?" I told the fifty, sixty guys there that I was quitting the association because they were a bunch of red-neck jerks. I came out of the meeting a director.

Pacific Trollers Association was not that politically minded back then, except on allocation issues. But I got very involved with the whole environmental end of things. I would go to other environmental meetings—the Save Howe Sound Association, Greenpeace, West Coast Environmental Law—and bring the information back to our association.

I felt that the fishing industry has been really, really good to our whole family, but I realized that you can't just keep taking. You've got to put something back in. I wanted to get people thinking about habitat protection and awareness, so I'd go to the PTA meetings and say to the guys, "You're always fighting over how to divide up the pie, but no one's worrying about taking care of the ingredients that go into the pie. We gotta keep growing those apples, guys."

It was really a tough go. These guys just didn't want to hear it in the beginning, but that's changing slowly. Taking care of the resource has got to come first. Not the money part of it, but taking care of the coast, habitat protection, estuary inventory programs. And it's not just runs and run sizes, but it's also stock diversity, keeping each stock strong. If our coastline and industry is dying, we need to give it CPR—conservation, protection and restoration.

My kids think my political involvement is fine, but if I get really involved in something they remind me that it's my trip. They can get really fed up sometimes when I'm in a big flip out, but it's worth it to me when people become aware, when organizations form, when things like the high seas drift net issue can go from concern in small coastal communities to the United Nations in seventeen months. Once you've been there and seen the situation for yourself, you have a different level of commitment.

People still think what has happened on the east coast will never happen to us. We're all a little too comfortable. But to make it work, all of us—the sports guys, natives, and the commercial fleet—are going to have to take a cut. We've got this jewel here. We're all going to have to work to keep it. 🐟

We started building 'Ksan in our yard. People would ask, "What are you doing?" We'd just tell them we were building a Viking ship. They'd believe it, it was so huge. Photo: Lemmers collection.

GILLNETTING
FOR
SALMON

Jarvis's original drawings as submitted to the patent office in the 1930s. Photo: D. Hufnagel collection.

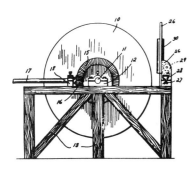

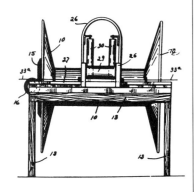

SOINTULA
Birthplace of the Gillnet Drum

Richard Gross

Today on a modern gillnetter, the fisherman has only to place his foot on the treadle that controls hydraulically run gears to haul his net. As in most technological innovations, underlying this engineering simplicity was a bright idea followed by a long period of trial and error to make the concept a reality.

This first gillnet drum led the way for the seine drum and similar devices that immeasurably eased the fisherman's toil, and increased the catching power of fishing fleets around the world. Ironically, the inventor himself didn't make a cent from his ocean-shaking concept.

Laurie Mattias Jarvelainen was nine years old in 1901 when he was sent from Helsinki, Finland to be raised by his aunt and uncle on Sointula. He was among the first wave of sturdy Finnish immigrants to settle the planned utopian community on Malcolm Island, off Vancouver Island's northeast coast. Although Sointula was founded as an agricultural community, its residents soon turned to logging and fishing for their livelihoods.

In his early teens, Laurie took to the loggers' trade, and between jobs he hoboed around the province. He changed his name to Jarvis, for most of his fellow loggers and travelling buddies couldn't pronounce the long Finnish handle.

He abandoned logging for gillnetting after he met his future wife, Helmi Johanna Lapti, at a Finnish dance in Vancouver. She had been born and raised on Douglas Island, Alaska, where her father worked a gold mine.

Jarvis joined the Sointula gillnet fleet of small, flat-bottomed rowing skiffs. There was only a tiny canvas tent at the bow to protect fishermen from the elements and serve as a primitive galley. It was a gruelling task to pull 200-fathom nets loaded with fish aboard these skimpy skiffs. Hardly a gillnetter escaped the pain of arthritic hands. Sometimes older fishermen had to wait for hours until their stiff fingers uncurled. Power boats were

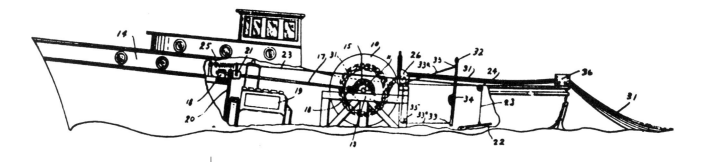

Before the advent of the gillnet drum, nets were hauled aboard by hand. Shown here are early 1900s gillnetters. Photo: D. Hufnagel collection.

introduced in the twenties and thirties, but nets were still tediously set and hauled by hand. Fishermen still had to anchor their boats in protected coves awaiting slack tides before they could set their nets. Bad weather also prevented them from fishing risky areas. There was just too much down-time for an ambitious highliner like Laurie Jarvis.

From his logging days, Jarvis displayed an imaginative, inventive mind, combined with resourcefulness and a natural mechanical inclination. Jarvis built and operated a small boatyard on the Sointula waterfront, and he soon made his name as a boatbuilder and troubleshooter. If a fisherman had a boat problem, Laurie was the man to see.

Jarvis continued fishing in season and the problems of hand-operated nets still plagued him. Wire rope used in logging comes wound on a heavy wooden spool up to five feet in diameter, and it was while he was looking at one of these castoff spools that the lightbulb of invention clicked on. Why not reel in the net the same way loggers wound cable on a wooden spool?

In 1931 Jarvis designed and built his first drum from local yellow cedar. Lauri Wilman, a fellow Sointula gillnetter, helped Jarvis with his project and was one of the first to install the new drum on his boat. Like most inventions that look simple, the drum presented a series of complex problems.

The first drums were too large and their cores were too low. The inventors turned to Wilman's wife Helen, a mathematics whiz, to calculate the proper size of the drum and core for the rolled volume of net. After months of experimentation and a half-dozen drums, the size was finally right.

Then there was the net itself to contend with. Years later Charlie Peterson, an eighty-eight-year-old retired gillnetter, recalled watching Jarvis making his first set off his new drum at Rivers Inlet. "The net snarled up so

Sointula

Laurie Jarvis had an imaginative, inventive mind. Photo: D. Hufnagel

Glen Hufnagel, grandson of Laurie Jarvis, next to an original cedar drum still in use in 1988. Photo: Richard Gross.

bad that they had to take it off [the drum] with a derrick at Wadham's Cannery. No matter how slowly and carefully they reeled in the net, the same thing always happened, and back and forth to Wadham's they went. It took the derrick and four men to undo the mess." Lauri Wilman discovered the trouble. On hand-set nets, efficiency dictated that the lead line be shorter than the cork line, but on the drum-reeled net, the short lead line was tangled by the longer, billowing cork line. "Lengthening the lead line was the winner," chortled Burt Peterson, retired skipper of the gillnetter *Ocean Dawn*, and the first person ever to reel a gillnet onto the drum. Perhaps the worst puzzle was the gears that drove the chain. Peterson remembers Jarvis and his own father experimenting endlessly with gear systems. "They tried a dry friction drive. That burned out really quickly. Then they installed open bevel gears from the rear end of a Model T Ford, but there was too much load for smooth operation. Next were heavy-duty worm gears off a truck, but they were too hard backing up. Finally, they installed the rear end gear system off a Model A Ford and that worked just fine."

It took more than a year to solve the riddle of the drum, but at last it worked and tested well at Rivers Inlet, and even in the tricky waters of Skookumchuck Narrows. At last, gillnetters would no longer be victimized by the tides and bad weather. Now, difficult fishing spots opened up to salmon-hungry gillnetters who could set and haul rapidly with the help of Jarvis's motorized, chain-driven drum.

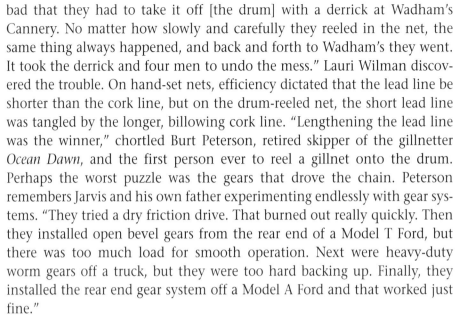

Visions of sudden wealth danced in Laurie Jarvis's head. He secured a patent on the drum, and according to his daughter, Sointula resident Diane Hufnagel, "Then the trouble started." First Jarvis went to a manufacturer, but their offer was so low that Jarvis converted his boatyard into a drum manufacturing plant. So many orders flowed in that the small plant couldn't handle them. It was the thirties and the depression was in full swing. To build the drum was relatively simple and the materials were readily available. Gillnetters from Steveston and elsewhere, either unwilling to pay Jarvis the $25 he charged or not prepared to wait for their orders to be filled, studied the drum in operation, went back home and built their own versions. Also, Sointula was historically a co-operative community, and its members frowned on profit-making by a fellow islander. Feuds erupted and Jarvis hired a lawyer. Judgements went in his favour, but at best all Jarvis could do was to force fishermen to remove the drums from their boats.

Modern aluminum gillnet drums powered by hydraulics. Photo: *Westcoast Fisherman* collection.

Fundamentally, a drum was only a large thread spool, and the patent for that had been secured more than a hundred years before Jarvis's time. The concept was almost too simple to make his patent stick. Suing individual fishermen for this infringement seemed like robbery, and legal fees were driving him to the poorhouse. After a few years of battling, he dissolved the patent.

Frustrated, Jarvis got a patent for another invention, a cradle-shaped pair of stern net rollers. They didn't work well, nor did they have the impact on the industry the drum had. Jarvis became embittered, but he kept fishing until his death in 1965.

The significance of Laurie Jarvis's drum to BC, in fact to the world fishing industry, is evident everywhere. So the next time you pull on the gear stick or step on a treadle to set or haul your net, give a thought to one of the unsung heroes of fishing, Laurie Jarvis of Sointula, the industrious Finn who made easy what was once a cold, wet, and backbreaking job.

Freshwater Bay

Billy Proctor fishing near Flower Island in his first fish boat. Photo: Proctor collection.

REMEMBERING PROCTOR'S FISH CAMP

Billy Proctor, as told to Alexandra Morton

The two-cylinder Atlas seemed to barely hold its own against the fury of wind and snow as the packer *Cheney* headed into the mainland, bound for Kingcome Inlet. On board was a young bride, Janet Proctor, on her way to Charles Creek to join her husband William Proctor. The skipper of the *Cheney* decided it was useless to continue deeper into the inlets in such a storm, so he guided his vessel into Scott Cove on Gilford Island and dropped the anchor. As the winds howled, Jae, as her friends called her, sat down to dinner, a can of pork and beans. It was Christmas Eve 1925.

William and Jae had married the year before. Shortly after their wedding, Bill took his pile driver up to Charles Creek to build a cannery for Mr. Stumpf of Kingcome Packers. When Jae arrived this Christmas day, her home was now a ten-by-twelve-foot shack built on an old scow. After two cold and wet winter months, she fled back to Vancouver. The following spring Jae returned and she and Bill moved into a company house on shore. Bill completed the cannery in a year and went on to build another at Leroy Bay in Smith's Inlet.

Late in 1925, Bill and Jae decided to try their hand at fishing. Each Sunday, a packer towed their little skiff along with nine others out into Smith's Inlet. At 1800 hours, they set their gillnet and fished steadily until 1800 hours on Thursday, when the packer returned for them.

Kingcome Packers gave their fishermen a boat and a net. The boats were 30 to 32 feet long, flat bottomed, with no mast. They were built with six shiplapped cedar planks and painted cannery red. These miserable little boats weren't named, just numbered, and powered by muscles and oars. There were two kinds of nets, a 5-inch net for sockeye and a 6½-inch mesh for coho.

The living quarters aboard these skiffs were sparse and primitive. There was a 6-foot piece of heavy canvas that could be secured across the bow,

Gillnetters being towed out to the fishing grounds—Kingcome Inlet, 1928. Photo. Jae Proctor.

and that was for protection against weather. The tarp was rolled up during the day to allow for more working room and it is hard to imagine that anything stayed dry for very long. Jae and Bill ate bacon and eggs, canned goods, and lots of stew. In addition, Jae used a four-gallon can on top of the stove as an oven to bake cookies. These warm, fresh cookies were their only luxury.

The Proctors began fishing Smith's Inlet for sockeye in 1927. In the first week of June, Kingcome Packers would tow sixty skiffs north, twenty at a time. During the trip up in 1929, Jae and Bill were aboard the *Saugeen* as she towed twenty little skiffs in her wake, looking like a mother duck with her brood. Jae had carefully stored all their belongings in their skiff, under the tarp. As the *Saugeen* crossed the mouth of Slingsby Channel, a westerly squall engulfed them. The wind and strong current combined to create a wild sea. Ten skiffs rolled, including the Proctors', dumping all their belongings into the seething waters. Their most precious possession was lost, Jae's camera. She cried for two days.

By 1929, the Proctors had saved enough to purchase the *Helga Herman*, a 32-foot gillnetter powered by a one-cylinder Frisbee gas engine. They renamed her the *Zev*. Jae was happy aboard her new home, and in later years she always remembered the boat as being very cozy. It must have been a huge relief not to live huddled under a canvas tarp for five months of each year.

Bill and Jae continued to fish Smith's Inlet for sockeye in July and then returned to Kingcome Inlet for coho and pink salmon. During the winters, they watched camp at the Charles Creek Cannery. Food came via the Union Steamships, along with the mail and 100-pound sacks of coal to heat the houses.

After two years of gillnetting on the *Zev*, Bill and Jae took the fish patrol job for Bond and Thompson sounds in Tribune Channel. They

Billy Proctor with a couple of Blackfish Sound springs. Photo: Jae Proctor collection.

walked the rivers, checked on the number of fish returning to spawn and patrolled the boundaries. In 1934, they took the fish patrol position for Port Neville and the Adam River. In Port Neville, Jae gave birth to their third child, Billy Proctor, who grew up to skipper the *Twilight Rock*.

In November 1934, Jae and Bill bought a beautiful little piece of land on Freshwater Bay at the north end of Blackney Pass. The purchase included a post office, store and fish-buying camp. The whole outfit, known as the White Beach Trading Co., sold for about $1,000. They changed the name to Proctor's Fish Camp.

Bill Proctor Sr. was doing well—too well in fact. He would go out for the day, load up on large spring salmon, and then return home and buy fish. The old-timers fishing for his camp gave him such a bad time about a fish-buyer catching all the fish that, after a year, he stripped the Zev and sold the gear for $10. The Proctors had plenty of fish-buying competition in their area. There was a camp in Baronet Passage, Mike Davis bought fish in Double Bay, Al Hood was in Yokohama Bay, and Spence Turner was in Mitchell Bay. They couldn't afford to drive any fishermen away. The Proctors sold the fish they bought to ABC Packers at Glendale Cove in Knight Inlet. ABC canned most of their fish, but salted the dogs and white springs.

In January 1942, Jae and Bill suffered a major financial setback when twenty-nine Japanese-Canadian fishermen who sold to them were imprisoned in the internment camps. But they had enough loyal customers to keep them going, because they treated the fishermen well. They gave them free moorage, and Bill took orders for meat and ran into Alert Bay once a week to pick it up. Fish prices that year went as follows: white spring, 2 cents a pound; red spring, 7 cents a pound; coho, 5 cents a pound; chum, 25 cents each; and pink, 5 cents each. Gas at the time was 19 cents a gallon.

On February 14, 1942, Bill Proctor went on a mail run to Alert Bay and never returned. He was last seen leaving the Bay at 2300 hours during a southeasterly blow. His boat was later found off the eastern end of Cormorant Island. For three days, seven-year-old Billy waited for his father to return. He generally accompanied his dad on these trips to town, but this time he had been left behind. He kept watch from a small island outside of Freshwater Bay, until finally the provincial police boat came out to give Jae the devastating news.

Jae must have been tempted to give it all up. In 1939 her little daughter Patricia had died, and she also lost a baby son. Now her husband

REMEMBERING PROCTOR'S FISH CAMP

The post office, general store and Jae and Billy Proctor's homo at Froshwator Bay. Photo: Jae Proctor.

Freshwater Bay, 1950. Photo: Charlie Gallenger.

Billy Proctor lands a salmon. Photo: Alexandra Morton.

was gone. Freshwater Bay had not been good to her. However, the sturdy Scottish redhead was not about to retreat to the city. Despite the fact that they were left with no firewood and no boat, she never mentioned leaving to her son Billy.

That winter brought three feet of snow. Their house was not insulated and the water line froze solid—all the water had to be packed in with buckets. In front of the house, they had an A-frame float that sat on the beach at low tide. Firewood was so scarce that Billy and Jae split the bark off the float logs and dragged it up to the house to burn, with mussels and all still attached.

Billy combed the beach for scraps of wood to burn and to repair the floats. When their closest neighbours, Bob Davis and Rod Williams, went to town, they stopped at the entrance of the bay. If Jae had hung a white cloth out on the clothesline, they would wait for Billy to row out to them with $10 and a grocery list. In the evening, they would stop again, and Billy would row out to pick up the grub.

Old Pearly Sherdahl, who owned the Minstrel Island Hotel, towed logs over for Billy and Jae to handsaw into firewood, and every Sunday he stopped in to check on them. In addition, he salvaged the *Zev*, patched her up and put her safely on the beach for Jae to sell. She got $65.

From 1942 to 1962, Jae ran Proctor's Camp on her own. The first few years were hard, but quite a few boats continued to sell to her. She paid cash every night, and gradually her reputation as a fair fish-buyer grew. When fishermen first arrived in the spring she would give them fuel and grub on credit if they needed it. Most fishermen were too independent to accept, but it contributed to her good reputation. On the other hand, if anyone suspected her scales were wrong or accused her of offering too low a price, she told them there was another camp in the next bay and they could sell there. Jae was tough, well respected and never backed down.

REMEMBERING PROCTOR'S FISH CAMP

During one particularly good spell of fishing, she ran out of cash. Besides the twenty-five steady fishermen who sold to her daily, another twenty-five appeared. Although she needed the business, she simply did not have the cash on hand. One of her steady fishermen offered to loan her the huge sum of $500 so she would not lose the newcomers. Jae hated to borrow money and in the end she didn't have to. A group of highliners dubbed the "Nanaimo Gang" told her not to bother paying them every night. At one time they trusted her with $3,000. These men had gained their highliner status by fishing deep and working at it all day, instead of quitting at noon if fishing was poor, as some of the old-timers did. They worked hard for their money, and their willingness to bank $3,000 with Jae was a real credit to her name. Some of these men continued to fish for her for twenty years.

Jae Proctor slinging fish. Photo: Proctor collection.

Her attitude must have been pretty no-nonsense. In twenty years of living and working in a remote bay, she experienced only one threatening situation. A fellow who had taken up drinking the alcohol used in stoves, came into the store wanting to buy all her stove alcohol. When she refused to sell him more than one can, he leaped over the counter, threw her to the floor, and grabbed all the remaining cans. He headed out the door and never came back.

Jae always kept a garden full of flowers. She particularly loved lilacs and honeysuckle. Inside the house, there was always a canary singing and a gold-fish bowl. Through the tough times, she refused to show any sign of being poor; she never ate broken cookies or allowed her son to go to the store with torn clothing. She described her life in poems that she carefully recorded in several volumes, but she never remarried.

In 1956, Jae decided that her business was suffering because of her location in Freshwater Bay. It was a terrible anchorage, offering minimal shelter from either southeasterly or westerly winds. When it blew, all her fishermen had to anchor on the other side of White Beach Pass, in Farewell Harbour. In addition to this, the thirty-two Nanaimo highliners were now anchoring on the north side of Swanson Island in Yokohama Bay. Spence Turner was sending the *Mitchell Bay* out to collect fish from them every night. Determined to regain these boats, Jae asked them whether they would sell to her again if she moved onto a float and tied in Yokohama Bay. They agreed, so she decided that now was the time to escape all the sad memories that had collected in Freshwater Bay.

With great effort, the building that contained her home, the store and the post office was dragged off its foundation and onto a float. Although the house was built in 1919, it survived the ordeal.

Floating unsettled Jae at first and she didn't care much for it. But the increase in their business was incredible. She was defiantly in the hot spot. Proctor's Camp operated for seven summers in Yokohama Bay. During the winters, Jae and her son moved the float to wherever Billy was hand-logging.

The worst year ever for fishing salmon in Blackfish Sound was 1962. High boat that year made $1,500. The skipper of that lucky boat, the *Twilight Rock*, was Bill Proctor, Jr. Taught by men like Finlander and John the Curd (John Kallas on the *Gurd*), Billy, like his dad, was found to have a special touch with spring salmon. However, in 1962 no one could make a living fishing in Blackfish. There were many closures to protect the few remaining fish and most fishermen moved west, into more open and productive waters.

On October 17, 1962, Proctor's Fish Camp moved out of Yokohama for the last time. Jae's little house moved in and around the inlets with Billy and his wife Yvonne until 1983, when it was moved back onto land. Billy and Yvonne bought land near Echo Bay in Simoom Sound and carved out a beautiful homestead there. Jae lived there only a short while, until she passed away in Alert Bay in December 1984. Jae Proctor was a remarkable woman, who will be remembered for tenacity and courage. ❧

LANDLOCKED GILLNETTERS
Fishing on the Stikine

story and photos by Dave Gordon

Some two hundred miles north of Prince Rupert, the Stikine River meets the ocean after a wild and torrential journey through the Coast Mountains. Its beginnings are inauspicious enough—a small glacial stream trickling into Tuaton Lake high on the Spatsizi Plateau. Fed by massive ice packs that are broken only by even more impressive mountains, the Stikine gains strength and momentum until it crashes through what is known as the Grand Canyon. Here it narrows to as little as 6 feet across in one spot. This 75-mile section of river is completely unnavigable. Fish are unable to get up any further to spawn, and boats cannot pass either upstream or down—though some people have died trying.

As it bursts out of the canyon, the Stikine becomes a very different river. Leaving the incredibly rugged coastal mountains for the mosquito-infested rainforest, it boils, twists and meanders in a fashion that is anything but gentle. It is in this lower section that a handful of fishermen try to eke out a living on the "Great River."

Their livelihood began with a test fishery in 1978. A year later, the federal Department of Fisheries and Oceans (DFO) opened the area to anyone who was willing to give it a go. Some of the sixty people who did were

Tahltan fishermen on the "Great River."

Dave Gordon

Two fishermen scoot by the "boundary house," which marks the Alaska/BC border on the Stikine.

Bob and Celine Gould and their daughter Jennifer fish from a Great Glacier Salmon drift boat, designed and built for the Stikine.

LANDLOCKED GILLNETTERS

moderately successful. BC Packers was on the scene, paying one dollar a pound cash to anyone who could deliver fish. The investment was relatively small, compared to salmon fishing elsewhere on the coast; a skiff, an outboard, and a net were all you really needed. But you had to get them there; build yourself a camp out of the wilderness; learn to navigate a boat on a wild river without hitting logs, sandbars or icebergs; keep your outboard running (with the nearest outboard shop a 24-mile journey downriver to Wrangell, Alaska); and, for many, learn how to fish.

DFO regulations require that a licence must be fished to be renewed the following season. The odds against success on the Stikine were so great that soon the number of active licences dwindled to a handful. BC Packers only stuck it out for two years before deciding that their efforts were not worthwhile.

Most fishermen on the Stikine set and haul their nets by hand.

The fate of the fishery was in the hands of the fishermen. Many of them were homesteaders from the Glenora area near Telegraph Creek, and they desperately wanted to keep the fishery alive. The opportunity to make some cash harvesting salmon from the river that fed their valley was one they did not want to pass up. So in 1981, a group of thirteen fishermen formed a co-op known as Great Glacier Salmon, named after a huge glacier that dominates the landscape on the lower river. They were a mixed lot: a few homesteaders from Glenora, a Tahltan from Telegraph Creek, and people from Prince Rupert and as far away as Montreal. They fished hard each season, pooling their resources and talents and any money made from the fish, and funnelled it all back into building a processing plant that could freeze their fish for marketing.

While they were successful in building their plant, the operation was only marginally lucrative, and co-op members as well as other fishermen left the river each fall with not much more than unemployment insurance stamps. The lack of financial reward, coupled with personality conflicts within the co-op and among other fishing groups on the river, led many fishermen to give up trying to make a living from commercial fishing on the Stikine. In the late 1980s, many were returning to the river each year just to make a landing in order to keep their licences.

At one point, Great Glacier Salmon fished with six licences, froze their catch and sold it to brokers for the best price they could get. A few miles upriver, the Tahltan Indians set up their own facility and fished with about six licences as well. They iced their fish and during the 1990 season they were able to sell some to processors in Alaska. This was more cost-effective than the previous system of flying fresh fish out in an Otter to Bob Quinn Lake, where a waiting freezer truck would transport them to Prince Rupert. Licences on the river are mostly held by local people from the Telegraph Creek area. While some fish fairly regularly each season, most have full-time jobs elsewhere and only turn up once a season. Fishing the Stikine has never been a way to make a fortune. If the year's total allowable catch for sockeye is 16,000 pieces and there are 21 licences, there is not much to go around.

The short history of this fishery has seen a staggering variety of boats and burned out a staggering number of outboard motors. John Plummer fished the river for years in a 16-foot riveted aluminum river scow with a 70-horse outboard on the stern. Loaded with fish, there was very little freeboard, and when the boat came off plane, the wake would come crashing over the stern, setting the deckhand to work furiously with the bailing can. Alan Reimer used to fish out of a skiff with twin 20-horses on the back. The

A typical fisherman's camp along the river. This 9 x 12 tent was home to John Plummer's family of four for four months.

sound of the engines whining in unison told everyone that Al was coming, long before he rounded the bend in the river. Dave Tauber used his 16-foot skiff *Butze* for years before it succumbed to hard work and loose rivets. Dave was a gifted outboard mechanic and managed to keep his 50-horse Mercury running for years, but others did not have the same luck. At the end of one season, John Plummer had five dead outboards sitting in camp and would pick his net from his engineless skiff and float down to the plant with his catch. Once, Dan Pakula and David Fisher had a brand new 20-horse bolted to a board nailed to two trees at the river's edge. They were proud of the fact that they were well prepared with a second motor—until they got up one morning to discover that the trees had fallen into the river with the engine and had disappeared.

Today, most of the boats are welded aluminum. Great Glacier Salmon went as far as custom-building a fleet of six boats specifically designed for the Stikine. This is basically a modified herring skiff with a Volvo Penta stern drive, and a power drum that reels the net aboard through a chute in the bow.

Bill Sampson worked with a custom-designed, 20-foot welded aluminum boat. Eight feet wide and powered by two 150-horse Mercury outboards, it could plane with thirteen full barrels of gas on board. Needless to say, his fuel consumption was high; he ran his gas line straight from a 45-gallon drum. Bill did most of his fishing out of a smaller skiff and chartered his big boat for fish packing, among other things.

There are basically two ways to gillnet salmon on the Stikine: set net and drift net. Set netting is familiar to many who have seen the native fisheries on the Skeena and Fraser rivers. The net is securely fastened to shore and set in a back-eddy in the river. The eddy keeps the net pushed out into the

Pacific Spray spent a couple of years on the Stikine, working as an all-purpose tender and packer.

river and the web hangs loosely in the area between the opposing currents. One anchor holds the net well out in the current and another keeps the tail end from folding back into the set. The result is a net that looks somewhat like the letter L, with the long axis pointing downstream and waving gently back and forth in the current. Until recently, this was the more popular method of fishing the river. Each fisherman maintained an unwritten ownership of his own eddy or "hole" throughout the season, although the river's constant changes would make fishing holes appear and disappear as the river meandered back and forth.

The fishery's boundaries extend from the Alaska border, about thirty miles from the river's mouth near Wrangell, upstream about thirty-five miles to Sterling Creek. Fishing is also permitted about ten miles up the

Dave Gordon

Iskut River—a tributary of the Stikine, to the infamous hole called "The Eye of the Storm" or "The Whirlpool." Here, the Iskut roars through a chute about 20 feet wide. Downstream on either side are two ferocious eddies that provide some of the best fishing on the river, if you are up to the challenge. Stefan Jacob recalls anchoring a brand-new, 100-foot coho net there with a 24-inch pink plastic buoy on the end. Shortly after setting the net, he and his deckhand pulled back his skiff to examine the set. To their horror, a whirlpool formed in the eddy, picked up a bit of momentum, and drifted towards their net, sucking all 100 feet of it and the buoy to the bottom. They landed their boat on shore and went to the tree where they tied off the shore end of the net only to find that the lead line and corks were tightly twisted around each other, and that the net had taken a sharp dive to the bottom and wrapped itself around something large and immoveable. Their only choice was to cut the net off and head back to camp for another one.

While this hole can be a dangerous and expensive one to fish, it can also be lucrative. Stefan averaged about 20 coho an hour when the fishing was good, and with set-netting, you leave the net in around the clock, so these numbers add up. There is a catch to this hole, though. The current is so strong there that the fish are "blown" out of the net almost as fast as they go in, so you have to be right on top of it and pick the net frequently. As well as some pretty big salmon, some of the fishermen have caught rocks as big as a human head.

While some fishing holes have names that indicate their character, such as "Fast Eddy" or "The Whirlpool," others are named after geographical or historical landmarks. The "Boundary House" hole is just upstream from the oldest building on the lower river—the boundary house, built as a border house early in the century and still standing—and has been fished for years by the Great Glacier Salmon fishing team of Bob and Celine Gould. The fishing hole that bears its name is one of the top producers on the river, but is also one of the most tedious ones to fish, due to the incredible amounts of flotsam and jetsam that circle through the eddy. This might be one of the worst drawbacks of river fishing. Every time the river goes up, it picks up sticks, branches and logs and sends them downriver. On their way, they pass through the back-eddies, but not through the gillnets. This is especially bad during the coho runs in the fall. A good rain can send the river up ten feet or more in a day, and in no time your gillnet is completely tangled in a mess of branches and leaves. Dave Tauber and Bob Warren fished the hole during one high water, and were pulling fifty or so coho out each time they picked their net. In the

John Plummer takes a moment to relax in his 16-foot skiff after a good night's fishing on springs and sockeye.

LANDLOCKED GILLNETTERS

As the cork line shows, Dave Tauber and Bob Warren have plenty of back eddies to contend with as they pick their net.

morning, they motored up to the set, hoping to load up on the 12- to 15-pounders they had caught the day before, but instead were greeted by a net so clogged with logs and sticks, they got out of their boat and walked from one end of the net to the other. They spent the rest of the day trying to salvage what they could, picking out the debris.

The "Pohle Lumber" hole is named after one of the many logging companies that have tried to harvest the Stikine's prime timber. Like all the other companies, Pohle Lumber lasted only a short while before giving in to the odds, but its name lives on among the fishermen. This hole is renowned for its seemingly endless supply of spring salmon. Dan Pakula and David Fisher set a pair of 75-foot nets there for the first opening of the 1982 season. By the time the fishery closed, twelve hours later, they had hauled 110 big springs into their 26-foot river boat. In subsequent years, DFO limited fishing in this hole until the springs were well on their way upriver, but it is still known as one of the best spots for springs on the Stikine.

While set netting used to be the most popular way of fishing the river, gradually the emphasis changed to drift netting. Great Glacier Salmon designed their fleet of aluminum boats with drift netting in mind. The boat's hydraulic drum could quickly bring a net aboard with minimum effort from the fisherman, and if you are drifting toward a log jam at 10 knots, you want to be able to get your net aboard with no hassles.

Drift netting was done almost exclusively at a spot near the border called "Dream Eddy." Four-hundred-foot nets—the longest allowed on the river—were paid out in an arc across the river. The net was allowed to drift from ten to fifteen minutes, about half a mile downstream. Then it was hauled aboard either manually or mechanically, depending on the boat, and the fish were picked out. Twenty sockeye was considered a good catch. This sort of fishing went on around the clock, if the fisherman could stay awake.

Openings dwindled from five- and six-day-a-week openings in the early eighties, to twenty-four- and thirty-six-hour openings in the early nineties. Drift netting had been avoided in the past because of its terrible toll on outboards and nets. But as the fishermen gained more experience with the river and with fixing engines, drifting became a plausible alternative. It was usually much more lucrative than set netting, too, and with shorter and shorter openings, the fishermen had to find better ways to catch more fish. To simplify matters for the drifters, a few dedicated fishermen would arrive on the Stikine early in the spring, when the water was low, set out in a skiff and remove most of the major debris by dragging a chain across the river bottom. This paid off for everybody during the fishing season. A whole net can be trashed in a matter of seconds if you snag up. Or worse yet, the net can fold onto itself and get tangled in the propeller; this has sunk more than one boat. If the stern of a small river boat is caught facing upstream, the boat can disappear from under its skipper in less than five seconds, leaving him afloat in the middle of the river. This has happened to a few fishermen, but luckily no one has been hurt.

Going through a lot of web is a fact of life on the river. Snags, logs and even icebergs are caught on a regular basis. It didn't take long for fishermen to learn how to hang and repair nets. "We hung 15,000 feet of net, in 1988, and had nothing left at the end of the season," recalls Dave Tauber.

By the mid-nineties the Stikine salmon fishery had settled into a slow but steady rhythm. But the fishery was threatened by pressures from outside. The enormous hydro-electric potential of the Stikine and the Iskut was being hungrily eyed by power developers. Mining activity increased dramatically. Here as elsewhere in BC, fishermen were asking themselves just what the future might hold.

Looking for a good spot to set.

THREE GENERATIONS OF ANNIEVILLE FISHERMEN

Marjorie Simmins, photos courtesy Vestad collection

I t is a rainy, dark-skyed fall morning on the drive out to Everald and Mardell Vestad's hillside home above the Fraser River's Annieville Slough in North Delta. Fortunately, the kitchen and the company are as welcoming as a sheltered cove in a gale. At the comfortable kitchen table, nineteen-month-old Trevor is enjoying his breakfast of toast and jam while his father begins telling the history of Trevor's great-grandfather, Erling Vestad Sr., and with it, the story of a remarkable fishing family.

"My grandfather was a wonderful man," says third-generation fisherman Everald Vestad. "He taught us so much and always had time for each of us." Although the "us" Everald is referring to is his four brothers—Marvin, Clifford, Howard and Jim, all fishermen—it soon becomes obvious that this measure of attention and love was extended to all the members of Erling's family.

"Erling Vestad Sr. passed on a great heritage to his family," reads the letter from Everald's mother, Miriam, who is married to one of Erling's five children, Ole Vestad. "A heritage of knowledge, love and understanding, for which we are all grateful."

It's a long way from the tiny island of Sekken off the coast of Norway to the province of British Columbia in Canada. But, for Erling Vestad Sr., who undertook this journey in 1927, it was the beginning of a life of independence and adventure.

Erling Vestad began fishing in Norway as a deckhand aboard the sturdy cod boats that worked the waters around

Karen and Erling Vestad Sr. in the back yard of their Annieville home (late 1960s).

The Vestad family fleet ties up at Annieville Slough on the Fraser River.

the Lofotan Islands. In those days before quotas and short seasons, they worked for as long as it took to catch and salt down a full load of cod. Often this meant many weeks offshore.

"My grandfather used to joke that the skipper who first took him on as a deckhand hired him so he couldn't court his daughter, Jenny," says Everald with a smile.

As fate would have it, Jenny and Erling would share part of their lives together—but not until half a century after she first caught his eye.

In the meantime, Erling took a job at a shipyard in Vestness, where he gained the knowledge that would later enable him to become a versatile shipwright in Canada. It was on his way back and forth to work, by row-boat, that he stopped off at the village of Vikebukt and met Karen Julie Vike, who became his wife. Erling built a house for himself and his bride, and in time, two sons, Trygve and Ole, were born to the couple. It seemed as though Erling and Karen had all the elements for a happy life.

But economic conditions in Scandinavia were poor. The questing spirits of many young Norwegians looked over the Atlantic Ocean to the United States and Canada, to countries where the opportunities for work and advancement seemed more abundant. And so, in 1927, Erling travelled to the delta lands of Richmond and built another house. Two years later, he was joined by his wife and sons.

A new country, an old problem: the Vestads had emigrated smack-dab into the "hungry thirties." For a while Erling worked at a sawmill in Surrey, sometimes walking the five miles to work to save a nickel. In the middle of this difficult period he realized that, with the skills he had acquired in Norway, the solution to lean times was as close as his backyard, and Vestad Boatworks was born. The first of over one hundred Vestad vessels—which today ply the coast from Alaska to Washington State—would soon be launched.

THREE GENERATIONS OF ANNIEVILLE FISHERMEN

Erling, after returning to Norway. He died on the island of Sekken, within half a mile of his birthplace.

And how much did a skookum gillnetter fitted with a car engine sell for in the early thirties? Six hundred dollars, and worth every hard-earned penny, especially when you consider the effort of friends, family and neighbours who hauled the vessel by hand, across greased two-by-four planks, over the mile-long route from the boatyard to the Fraser River.

"Best of all," writes Miriam, "Vestad launchings always included coffee and Grandma Karen's famous banana cream cake."

Boats weren't the only special creations being christened in these years. Karen and Erling welcomed three more children into their lives. These were first-generation Canadians they named Lily, Stanley and Erling Jr. By now the family had moved their home to the slopes above Annieville Slough, where Erling continued his boatbuilding and repair business.

After sixty happy years of marriage, Erling Vestad found himself a widower in Canada. Seeking to lift his sad spirits, he travelled back to Norway, and there he kept a long-standing appointment with fortune: at eighty years old, Erling married seventy-eight-year-old Jenny, his childhood sweetheart. Together they returned briefly to Canada and spent their honeymoon fishing the waters of Rivers Inlet—a Viking honeymoon, to be sure.

"In Norway, at eighty-four he was still fishing from a small rowboat," says Everald. "His catch fed the village. Then one day he slipped on some kelp, took a knock on the head and died on the beach."

If there is a gene for producing children who call the ocean home, then the Vestad clan has it in abundance. Three generations of sons have become fishermen and a fourth generation is eyeing the prospects. The women, too, have been immersed in the industry. Erling's daughter Lily and her family manufacture sports tackle in Seattle. And Miriam, retired now from real estate and with all five sons grown, enjoys salmon fishing with Ole, who at sixty-nine years old, has no plans to retire.

Annieville Slough in the 1990s.

"I still get butterflies in my stomach at the start of the season," Ole says, during a telephone conversation from his and Miriam's winter home in Yuma, Arizona. "I love the freedom of fishing, being able to go where you want and do what you want. There's no reason to retire, if you still enjoy it. And I do." So do his brothers, Stanley and Erling Jr., who also remain active in the industry (oldest brother Trygve, passed away several years ago).

The lifestyle of commercial fishing—long hours spent on repairs and maintenance, and weeks and months spent on the fishing grounds—can be tough on families, and the Vestads are proud that they have maintained close family bonds.

"Our boys, no matter where they are or what they're doing, always send us Mother's and Father's Day cards," says Miriam about their sons. "There is a lot of respect and love between all of us." As for the successful rearing of five boys, Miriam downplays her hard work.

"You did what you had to do, and I was lucky enough to have very supportive parents and parents-in-law," she says.

Everald has equally warm words about his parents and siblings. During the salmon season, when the family fishes at Rivers Inlet, he and his brothers keep in close contact. "We always know where everyone is, and we'd be there in a flash if someone got in trouble."

Happily, the Norse gods have protected their own. None of the Vestads has been seriously injured or lost a vessel in the many years they've fished BC waters.

"I plan to build the rest of my life around fishing," says Everald. "The few times I've considered quitting, I couldn't think of anything else I'd rather do."

"It's the lifestyle that's so attractive," he continues. "We fish the salmon, herring and halibut seasons, and in between we're down at the boat shop in Annieville, maintaining the boats, canning salmon, and eating Dad's delicious fishcakes."

The last Vestad launch was in the late 1960s. Fishing, and looking after their own boats (there are eleven) and families keeps the Vestads busy and content.

Howard's son Bradley is surrounded by fish during the 1993 Fraser opening.

A FAST AND DANGEROUS RIVER
Gillnetting on the Skeena

Skeena River

story and photos by Peter A. Robson

The weathered orange Zodiac raced at 30 knots across the brown, choppy waters of the Skeena River on British Columbia's north coast. It was 1745 hours on a late July 1995 afternoon. In fifteen minutes the Skeena was to open to gillnetters for a forty-eight-hour fishery.

Burly Dave Lewis, dripping from spray, was hunched over the wheel of the rigid-hulled inflatable in his well-worn cruiser suit. Lewis owns the 60-foot *Princeton 1*, which is anchored nearby. Since 1990, he and his boat have spent salmon seasons on charter to the federal Department of Fisheries and Oceans.

Lewis monitors commercial fishing on the Skeena from its lower reaches at De Horsey and Kennedy islands to the up river fishing boundary at Mowitch Point. He provides the DFO with counts of gillnet vessels fishing

The Skeena River

Peter A. Robson

Frank Brown's *Captain's Dream* was fishing the eddy behind Veitch Point, known by locals as Lambert Point. Nearby, George Brown hauled his net aboard his *Ocean Freedom*. Just outside this eddy, the river rushed and swirled dangerously in its full ebb. Frank and George have a reputation as two of the Skeena's top highliners. George said this eddy can only be fished during very specific times and tides. He was born and raised at nearby Port Essington, long since abandoned, and has fished the Skeena and most other areas of the coast all his life. His grandfather and father were both Skeena River fishermen. When George's father lost his sight during his later years, he continued to go out fishing with his wife; she'd set the net on his instructions.

Phil Stevens aboard the *Margie*. Phil's been fishing commercially since the 1930s. What's kept him going? "Fishermen are dreamers," he smiled.

in the area, their estimated catches and the mix of species. He also makes sure no vessels set their nets early, leave nets in after the closing, or fish outside the boundaries. Lewis has an easy-going but firm manner in his dealings with Skeena fishermen, and his non-cop-like attitude has earned him a great deal of respect among the fleet.

The tide had been ebbing for an hour or so and about thirty vessels were waiting near the up river boundary for the signal to begin fishing. Lewis cut the throttle and the inflatable drifted. He watched for boats setting their nets early. Moments later, the opening announcement came over the VHF radio. Suddenly the fleet was in action and nets began streaming overboard. Lewis made a quick count of the vessels nearby and raced off to count the rest of the fleet in his patrol area.

The Skeena is the second largest river in BC, as well as the second largest producer of sockeye salmon—rivalled only by the Fraser. It is some two miles wide near its mouth and subject to tidal changes of up to 24 feet—an amazing rise or fall of a foot every fifteen minutes. In 1994, 2.3

(Below) Alfred Nellie and his grandson Dave Alfred of Alert Bay were fishing Hells Gate Slough aboard the *Rembrandt II* (left).

(Left) Andy Rosengarden aboard his *Lady Karen*. Andy was originally from California. He was working most of the year as a schoolteacher in Richmond, BC and fishing during summer break.

(Below) Harold Wulff aboard the *Joanie*. Harold has fished the Skeena since 1982. The difference between fishing the Skeena and the Fraser? The Skeena has: "Big tides, you can't fish the ebbs, lots of debris and big stumps." On navigation: "You might as well throw your charts away. The local names have nothing to do with the chart names."

Vince Muldoe and the *Daz'lin Pearl*.

Roy Wilson and his son Chad.

Chris Skulsh and his son John. Chris's dad, Roy, fished the Skeena under sail—before gillnetters had motors. Chris said his best year on the Skeena saw him catch 5,000 sockeye. His worst? 1960, when his season total was 128 sockeye. Chris had fished the Skeena and only the Skeena since the 1940s. He and his son were often spotted fishing the Glory Hole and Carlisle Bar.

million sockeye salmon were expected to return to the river to spawn. Commercial fishermen were to be allowed a catch of one million.

The river once boasted eighteen canneries including famous names such as Cassiar, Sunnyside, North Pacific, Balmoral, Inverness and three others at Port Essington. During the First World War, Britain was BC's biggest customer for canned salmon and the Skeena was where most of it was produced. After all, soldiers had to be fed well and Skeena sockeye are considered by many to have the highest quality of any sockeye. Today, the last of the canneries are gone, their sites marked by decayed pilings and a profusion of berries as Mother Nature reclaims the land.

The Skeena is a shallow river, where huge expanses of mud flats are revealed at low water. Along with its strong currents, this has always made the river difficult and dangerous for gillnet fishermen.

The combination has bred hard-working fish. During the ebb, returning salmon tend to hold in back-eddies, out of the main flow of the river. When they feel the push of the flood tide, they slip out of the eddies, up over the bars, and continue up river.

Just as the habits of the fish have changed little over the years, descendants of the pioneer European and native families set their nets in the same spots that

their ancestors did: the Glory Hole, the Boneyard, Lambert Point, Inverness Slough, Carlisle Bar, Blind Slough and the Long Nose, Mud Bay and De Horsey drifts.

At 2000 hours, Lewis arrived back at the *Princeton*. His deckhand Indira Persaud was busy cooking dinner. He used the VHF radio to contact the DFO office in Prince Rupert. Lewis reported that he had counted 200 gillnetters fishing in his patrol area and that the first set had provided an

Dean MacDonald caught this 60- to 70-pound spring salmon after it became entangled in his net. Dean broke a brand new gaff trying to bring the giant aboard.

Veteran Skeena fishermen Billy Kristmanson and his son Colin aboard the *Rival*. Originally from Iceland, the family homesteaded at Osland in De Horsey Passage in 1914.

Glory Hole fishermen Alex and Mike Morrison, aboard *Miss Jennifer Ann*. These guys handled their boat like professional race car drivers. They made numerous five-minute sets, with less than a quarter of their net out over the famous hole adjacent to Gregory Point. Then they drummed in at full speed. As soon as the flood rose a certain amount over the bar, they moved away from the spot and fished elswehere.

(Below and right) Linsey Husvik and son Lenny aboard the *Wind Walker* were fishing the Little Rip Drift off De Horsey Island. Linsey's mother was born at the Sunnyside Cannery and his grandfather Andy Wilson fished the Skeena in the sailboat days. At least seven of Andy's descendants were fishing the Skeena in 1994.

average of twenty to thirty fish per boat—slow fishing. However, because the fishery had opened during a falling tide, the fish weren't moving and good fishing wasn't expected until the tide change, just before midnight.

At 0530 hours the following day, the fishery had been open for almost twelve hours. Lewis and Persaud raced off to collect tallies. When they returned, they reported, as expected, that the fishing had improved during the flood tide. The average for the night was 130 to 150 sockeye—decent fishing. Only about sixty vessels were fishing. The balance had apparently moved on to try their luck elsewhere.

While there are probably as many strategies for catching Skeena sockeye as there are gillnet fishermen, Lewis explained a technique used on the main body of the river. The fisherman sets his net at the up river boundary about two and a half hours before the end of the ebb. The boat drifts down river several miles. When the tide turns to flood, the boat is carried

A FAST AND DANGEROUS RIVER

back up river to the upper boundary where the net is hauled. However, by this time, the river is often flowing too quickly to fish safely. Rather than risk tearing up his net, a fisherman is better off to find a spot out of the main body of current, anchor up, and catch a few hours of sleep.

Twenty-four hours later, with the fishery into its thirty-sixth hour, fishermen were reporting poor catches for the previous night—an average of 15 fish per boat. The fishery was scheduled to close at 1800 hours. In the early afternoon, however, a twenty-four-hour extension was announced because target catches were not being made. When the fishery closed the following day, it was clear that the expected numbers of sockeye had not shown. Most fishermen who stuck out the opening ended up with 275 to 500 fish, about half their usual average for late July. By the end of the 1994 season, the forecast run of 2.3 million fish had been downgraded to 1.5 million and the expected catch of one million was downgraded to 500,000.

The low returns on the Skeena in 1994 contributed to the state of gloom that enveloped the BC salmon fishery at the beginning of 1996, spurring the federal fisheries minister to drastic action, but just when everyone seemed ready to agree the country's second largest sockeye run was in crisis, it came up with a surprise. Close to 7 million sockeye returned, providing a bonanza for fishermen and exceeding all pre-season escapement goals for sockeye by more than double. It was just one more reminder that man has a long way to go before he can claim to understand what goes on in a great fishery like the Skeena.

DFO charter patrollers Dave Lewis and Indira Persaud.

Not everyone can fish the Glory Hole, as evidenced by this large chunk of torn net and a buoy that were on hung up on a rock.

Illustration by Alistair Anderson

SHE'S MY DECKHAND

Suzanne Gerard

"Hey, Ma!" called Bruce, waving from the deck of the *Provider*.

"Be right there," I muttered, eyes riveted on the next plank I wanted to plant my foot on. A wave hit the rolling wharf, throwing me sideways. Stretching my arms out for balance, I shuffled stiff-legged towards my son.

"Decided to give it a try eh?" grinned Bruce. "Way to go!"

I shrugged with exaggerated bravado. "Why not?" We both sniggered as Bruce turned his back to the morning sun and pointed into the middle of over one hundred fishing boats moored around the Prince Rupert cannery.

"Fritz is over there, loading groceries onto the *Maggie*." I caught the movement of a plastic grocery bag rising in the air.

"Blue hat, black beard?" I asked.

"That's him," confirmed Bruce, sitting down on the hatch cover to comfortably watch the proceedings.

SHE'S MY DECKHAND

Tentatively weaving around and over gear to be loaded onto the fish boats, I ventured close enough to read the cannery's insignia on Fritz's blue baseball cap. I blurted out, "Hear you need a deckhand." The grocery bag paused in mid-air as two surprised brown eyes in a wrinkled, weather-beaten face turned in the direction of the female voice addressing him. My five-foot, middle-aged frame was given a cursory glance and dismissed with a grunt as Fritz lowered the bag onto the deck of the boat. "I'm a hard worker and reliable," I continued.

Another bag rose in the air. "You ever been a deckhand before?"

Remembering why he had fired the last deckhand, I responded with, "I learn fast and I don't drink."

Lowering the bag, he slowly turned and demanded, "Can you cook?"

"I'm a great cook!" I announced.

I must have hit on the right combination of assets. He handed me the grocery bag, told me I could use the rain gear left behind by the previous deckhand and said we would be leaving for the next four-day opening in an hour.

Sitting on the side seat behind Fritz, I looked around the cabin, which reminded me of a small trailer I had once owned—compact.

"My bed and the head are behind the curtain. The seat you're sitting on is your bed. There's a sleeping bag and pillow under it. We're going to get ice, so get your gear on," said Fritz, circling around the other fish boats moored at the cannery.

I pulled on the rubber pants, assuring myself that if I gained an extra 200 pounds they would still fit, and tucked the legs into my boots. The jacket was extra roomy with sleeves that hung down to my knees. Donning my toque and gloves, I sat in readiness and waited; for what, I didn't know.

We cued up behind the other boats waiting for ice. I clung to the back of Fritz's seat as the boat rolled with the waves, and watched over his shoulder to see how the ice was loaded. The deckhand ahead of us slid the hatch cover off and disappeared down the hold. A hose the size of the hose behind my dryer was lowered from the cannery into the hold after him. His skipper signalled to the men above and a white cloud of frozen air billowed from the hold. Piece of cake, I thought to myself.

"Fill the back and side bins first, then half the centre and front bins," said Fritz.

This was Greek, until I got down there. I aimed the hose into the back bin. A roar of ice exploded from the hose, throwing me against the front bin. A frozen, white cloud engulfed me as ice erupted up the hold and ricocheted off the boards. Sliding around trying to get control of the hose, I heard Fritz yelling commands that I couldn't make out. The ice miraculously stopped. He appeared out of nowhere, yanked me up by the collar, grabbed the hose out of my arms and yelled, "Stand like this and aim!"

I noted his solid stance with the hose under one arm and his left hand over the lip of the hose directing the ice.

"Come here and grab it!"

I scrambled through the loose ice and did as instructed. Fritz signalled for ice and kept me upright with one hand pushing against my back. We loaded the bins and he signalled for the ice to be turned off.

"You got a hell of a mess to clean up here," he said, handing me the shovel.

My stomach felt queasy as I sat on the deck emptying ice out of my boots. The patch to prevent seasickness was stuck behind my ear, but hadn't been on long enough to have any effect. Until my stomach settled down, I stared at the stationary mountains that stood like sentinels along the channel.

"Wash the ice and salt off your gear before you come back in the cabin," bellowed Fritz.

I hung my jacket over the safety ropes, hauled up a bucket of water and poured. Knowing my rear was also covered, I dumped the next bucket over my shoulder. My rubber pants were immediately filled with ice-cold ocean. I groaned and shook my leg like a wet dog. Hoping Fritz hadn't heard or seen me I peered in his direction. A smirk stretched halfway up the side of his face.

"Hang your gear up and change. You can use my quarters."

Relaxing in the seat beside Fritz after the supper chores were done, he said, "All I want you to do is cook, clean, ice fish, steer in a straight line when I set the net, wake me up if we drift too close to the boundary marker, and don't go anywhere near my net."

"Okay," I said.

Fritz stood up and pointed to his seat. "Sit here and take the wheel." Cautiously, I slid into his seat.

"See that island dead ahead?"

I nodded.

"Aim at it."

Adjusting the speed, he turned and disappeared out the cabin door to set the net. Panic-stricken, I grabbed the wheel with both hands driving as though I were in a car, but there was no white line, no solid road, and no brake, just waves that seemed to push me sideways. What felt like an eternity passed as I frantically tried to stay on course.

"Move," came the command from behind me.

I scrambled out of his seat. He wheeled the boat around and followed the floats back to where we started.

"If any of the floats are pulled below the surface, we got fish in the net. Get your gear on."

Untangling a fish from the net, Fritz held it up by the gills and asked, "Know what this is?"

Annoyed at how dumb he thought I was, I answered, "A fish." He stared at me for a few seconds and then tossed the fish on the deck.

"A pink," he stated. It looked more green than pink. Another fish landed at my feet. "Coho," he announced.

Oh! Types of salmon, I thought, making the connection.

Fritz interrupted my studies of the fish with the directive, "Throw 'em in the hole."

SHE'S MY DECKHAND

I gently picked up the "coho" and stared at the eye that was staring back at me. There was intelligence there and I wanted to throw it back into the ocean.

"The hold!" roared Fritz.

I cringed as I let the first few slide into the hold. Some of the salmon got even with me. Being slippery, they wriggled out of my grasp or slapped me in the face with their tails. I thought this was fair.

I lost count of the number of times the net was set and emptied. Night had come and I hadn't noticed. Lights came on, including one that had been lowered into the hold.

"Ice 'em down," said Fritz, pointing down the hold.

I lowered myself down, trying to avoid slipping on the fish below my feet, and wondered if salmon bit. I picked up one close to me. It was stiff with dull eyes. Sadly, I dug a little grave for it in the ice and put other fish beside it in a row, adding the odd handful of ice here and there to make sure they were well covered up. Fritz's voice boomed down the hold.

"Ya' don't have to tuck 'em in for the night. Grab the shovel and ice 'em."

I complied.

Tired and aching all over, I flopped over the edge of the hold onto the deck, which was covered with fish, blood and slime. I didn't have the strength to fight with my stomach any more and vomited into the muck I was lying in. The retching stopped long enough for me to crawl over to the bucket and haul up some water, which I poured over myself. I hauled up another bucket of water and pushed it towards the fish. I felt miserable and exhausted as the retching continued. Dunking each fish in the bucket of water, I let them slip out of my hands down into the hold. The silence from Fritz was deafening as I dumped the bucket of water onto the deck and pushed the muck into the sea. I sat with my head resting against the boards wishing I was dead.

"Go take your gear off and lie down," Fritz said. "I'll call you in a few hours."

The next few days were devoted to fishing, eating, cat-napping, and learning. My stomach came to terms with the seasick patch and stayed down.

The last day out brought with it a blinding rainstorm. I had managed to make quick trips to the toilet when Fritz was not around, but this was a serious, solid, four-day retention problem. Fritz was sleeping with his head less than a foot from the head. Drifting with absolutely nothing in sight to hit, I donned my rain gear, grabbed a large, empty, Melitta extra-fine ground coffee can, and crept out the cabin door. Leaning into the wind and rain, I ducked under some hanging ropes. The task accomplished and clothing readjusted, I threw the can overboard, expecting it to sink. Squinting into the rain, I watched in horror as the can bobbed in the waves towards the net.

"Oh, no," I moaned, visualizing Fritz hauling in more than he bargained for. I searched frantically for something to throw at it. Grabbing a long aluminum pike pole, I stood poised to throw it like a harpoon, when the can bobbed up and over the net. Breathing a sigh of relief, I turned to put the pole back.

"Whale?" asked Fritz, arms folded, leaning against the door jamb.

Back at the cannery, after the fish had been unloaded, I swayed down the wharf towards Bruce's boat and tripped over some gear in my way. Fritz slammed me on the back as he strode by with one of his buddies.

"See you in the morning," he said.

Their conversation continued as they walked on. "She's my deckhand." 🐟

HURRICANE FORCE
The Storm of October 1984

compiled by A.J. Provan

Brooks Peninsula

On a fall day in 1984, a hurricane-force storm ravaged the BC coast. Distress calls flooded the airwaves from all around Vancouver Island, yet rescue crews were thwarted by the force of seas being propelled by winds gusting up to 80 knots. In the tempest's wake, many vessels capsized and spilled their crews into the icy water to perish.

The mercy of the sea can be as fickle as feline loyalty. Its nature is both benign and wrathful. The following account illustrate the sea's rage, and the price it extracts from its children. The drama is condensed from *Casualty Investigation Reports*, Numbers 310 and 312, issued by Marine Casualty Investigations, Transport Canada.

Thursday, October 11, 1984

1115 hours. *West coast Vancouver Island. Winds southeast, 10 to 20 knots today, increasing 20 to 30 Friday morning. Mostly cloudy with showers, chance of a thunderstorm today. A few morning fog patches. Rain beginning Friday near noon. Visibility, at times, three miles in precipitation and near zero in fog.*

There was nothing in the forecast to alarm the salmon fleet assembled on the west coast of Vancouver Island. There were 92 seiners and around 200 gillnetters in the vicinity of Nootka Sound, and 48 seiners and 30 gillnetters near Kyuquot Sound. The fishery had closed at 1000 hours on October 11, and most of the catch had been transferred to fish packers. The fleet was preparing to leave the area.

1500 hours. The *Lady Val II,* owned and operated by Ken Datwiler of Courtenay, sails from Nootka Sound in company with the *Silver Triton.* The two gillnet vessels proceed northwards at a speed of 7 knots in an easy sea with negligible wind. The *Silver Triton* is operated by Richard Cowlin of Prince Rupert with Patricia Maleshewsky as crew.

1900 hours. *Lady Val II* and *Silver Triton* are proceeding 5 miles north of Esperanza Inlet when a third fishing vessel, the gillnetter *Miss Robyn,* leaves the inlet and takes up position astern. John C. Secord, Jr. of Cortes Island is the owner and operator of the *Miss Robyn* with Thomas S. Szczuka as crew.

1915 hours. *West coast Vancouver Island. Gale warning issued. Winds southeast 10 to 20 knots tonight, increasing to 30 to 40 Friday afternoon...*

2130 hours. The *Silver Triton* and *Lady Val II* are now three miles offshore, to the south of Brooks Peninsula, with the *Miss Robyn* four miles astern. The vessels are encountering a moderate swell from the southeast with the wind also from the southeast at a speed of 20 knots or so. The *Silver Triton* and the *Lady Val II* slow down to allow the *Miss Robyn* to catch up, and by 2145, the wind speed has increased to an estimated 30 knots.

2300 hours. *Storm warning for west coast Vancouver Island... Winds are expected to increase to southeasterly 45 to 55 knots with higher gusts...*

The *Silver Triton* and *Lady Val II* are off Solander Island with *Miss Robyn* approximately one mile astern. All three vessels are experiencing rough seas and the wind has increased to an estimated 50 knots. All three attempt to seek shelter on the north side of Brooks Peninsula, but the *Lady Val II* is unable to turn in the severe sea and wind conditions. She is forced to proceed northwards, towards Winter Harbour. Ken Datwiler dons his immersion suit. After proceeding three or four miles to the north, he is able to come about, but he cannot make headway against the storm-force wind and 30- to 40-foot swells, and the vessel is again headed north.

Friday, October 12, 1984

0100 hours. Two seiners, the *Miss Joye* and the *Western Hunter*, catch up with the *Lady Val II* after homing on her searchlight beam. They reduce speed to provide an escort.

0227 hours. Alert Bay Coast Guard Radio Station receives a call from the *Miss Robyn* stating their position and advising "that the winds were very high and they were not able to make shelter, and that if the winds didn't get worse they would be okay."

0229 hours. The *Miss Robyn* reports that another vessel and itself are in trouble. The Rescue Co-ordination Centre in Victoria is alerted, and, at 0233 hours, an "Urgency" signal is transmitted on distress frequencies.

0253 hours. *Silver Triton* calls Alert Bay "asking for a Coast Guard vessel to give them information on the best way to go, that they were in trouble due to high winds, and that they needed help." The Canadian Coast Guard cutter *Ready,* at anchor in Esperanza Inlet, fifty miles to the southeast, is alerted and prepares to get underway.

0300 hours. *Lady Val II* and its escorts are now encountering high seas and winds estimated at 70 knots (hurricane force). Shortly afterwards, the *Miss Robyn* reports that they are in distress and putting on survival suits, and that they have not heard from the *Ready*. A "Mayday Relay" message is broadcast to alert all vessels in the vicinity, and at 0315 hours, the *Ready* is underway, followed by the fisheries patrol vessel *Tanu,* which had been secured alongside the town of Zeballos, in Nootka Sound.

0320 hours. Many vessels respond to the Mayday but are unable to assist. Only the deep-sea ship *Hoegh Marlin* is able to proceed towards the distressed vessels.

0424 hours. A garbled transmission is received from the *Miss Robyn* with the only discernible words being "breaking up."

0430 hours. Only one and a half miles southwest of Kains Island, at

HURRICANE FORCE

Illustration by Graham Wragg

the entrance to Winter Harbour, the *Lady Val II* is rolled on its side by the force of the wind and capsizes. The *Miss Joye* is close astern and the *Western Hunter* the same distance ahead. Ken Datwiler abandons his vessel wearing his immersion suit but does not have time to pull up the hood or close the zipper properly.

After an unsuccessful pass across the bow of the *Lady Val II*, the *Miss Joye* turns about and a life ring with line attached is thrown to Mr. Datwiler. As he is pulled alongside, the line either breaks, or comes loose, but Mr. Datwiler is able to resecure it. There is only one deckhand aboard the larger vessel and he is unable to lift Mr. Datwiler aboard in his partially flooded immersion suit. The skipper has to leave the controls to help get him on board. Mr. Datwiler is uninjured, but shocked, and the *Miss Joye* and the *Western Hunter* proceed through the high seas and storm-force wind to the safety of Winter Harbour.

0504 to 0630 hours. Several calls are heard from the *Miss Robyn* to the effect that "they didn't know if they could hold out much longer," and "they were taking water over the decks." Finally, at 0630 hours, assisting vessels received the message "going over, hope you can find us." The storm warning is continued in the forecast issued at 0515 hours, which predicts winds from the southeast at 45 to 65 knots, with higher gusts diminishing later in the day and shifting to the southwest.

0600 to 2000 hours. When abeam of Solander Island the *Ready* is encountering extremely rough seas with winds gusting to 80 knots and is forced to seek shelter in Klaskish Inlet. The *Tanu* reaches the designated search position at 0916 hours, and commences the search in a heavy sea with wind speeds exceeding 50 knots. The *Hoegh Marlin,* which is standing off the entrance to Brooks Bay, and two other deep-sea vessels that were proceeding to the scene are advised to stand down.

0930 hours. An Aurora fixed-wing aircraft from Comox is able to commence the search and shortly afterwards sights the wreckage of the *Silver Triton* with orange-coloured objects close alongside. While the *Tanu* is proceeding to the location the aircraft resumes its search, and at 1006 hours sights the *Miss Robyn* with the bow protruding from the water and the stern submerged. At 1038 hours, the *Tanu* identifies these orange-coloured objects as fishing floats.

1124 hours. The *Tanu* is forced to seek shelter in Quatsino Sound due to the extreme conditions, but resumes the search when the aircraft reports "what appeared to be two persons in the water." Further brief sightings are made, and, at 1250 hours, a second group of persons is sighted close to shore. The *Tanu* cannot safely approach the second group and continues towards the first sighting, searching unsuccessfully until darkness in virtually zero visibility due to the heavy seas and spray.

On learning of the sightings in the water, the pilot and engineer of the Labrador helicopter on standby at Comox start the engine in wind conditions exceeding the safe limit for engaging the rotors. The remainder of the crew are embarked and the aircraft proceeds to the scene of the distress. Unfortunately, the wind conditions prevent the helicopter from reaching the immediate area until the following day. Meanwhile, a party of eight volunteers from several vessels in Winter Harbour lands by the fishing vessel *Caamano Sound* above Kains Point. The shore party makes an arduous crossing of the headland in the hope that survivors may have been washed ashore; however, the search is curtailed by approaching darkness, and the party returns to their vessel without finding any trace of survivors. Five of the party are thrown in the sea when their skiff capsizes on its way back to the *Caamano Sound.* All five make it safely back to the shore and are eventually taken off by a boat from the *Tanu* at 1940 hours.

Saturday, October 13, 1984

The sea and wind have moderated by the morning of October 13, allowing the search to be resumed. At about 1100 hours, the bodies of Richard Cowlin and Patricia Maleshewsky are located on the shore, still in their survival suits, and, at 1600 hours, a technician lowered from the helicopter to the wreckage of the *Silver Triton* finds the body of Thomas Szczuka trapped inside. The operator of the *Silver Triton,* John C. Secord Jr., is never located, but on October 17, a survival suit identified as belonging to the *Silver Triton* is found in the vicinity of the wreckage.

Fatalities from the October 11 storm include Richard Cowlin, Patricia Maleshewsky, Thomas Szczuka, John C. Secord Jr. and Charles Casey, whose crab boat *Hurricane I* was dashed on the rocks off Victoria.

SEINING FOR SALMON

Previous page: seining in Johnstone Strait. Photo: Alexandra Morton.

ALWAYS HOME FOR CHRISTMAS
The Martinolich Family

Alan Haig-Brown

Glenn, Richard and Aldwin Martinolich, 1990, in front of a modern version of the drum they helped develop. The Martinolich family were pioneers in the development of the BC seine fishery.

If you mention seining and Ladner in the same breath on the BC coast, people just naturally expect to hear the name Martinolich. The family has been in Ladner since before the turn of the century and they were among the first to bring purse seining technology into Canada from Washington State. Their boat the *Yankee Boy* was one of the first seine boats in the province when Mat Martinolich brought her up from the states in about 1911. When he built the *BC Kid* two years later, she was one of the first seine boats built in BC. (She was rebuilt and renamed the *Homelite* by Bill Warnock and in 1996 was still fishing as a troller under Derek Prince.)

Mat's sons Aldwin, Glenn and Richard Martinolich have been industry pioneers in their own right. They took the budding drum seine idea in the early 1950s, built the *Mar-Lady*, *Marsons* and *Mar-brothers* and developed the drive systems and nets to make drum seining practical. Every seine fisherman who hasn't had to pull corks in the skiff in a Johnstone Straits westerly owes these guys one.

For a hard-fishing family like the Martinoliches, Christmas was always a special time. Richard remembers it especially, as a break in the hard and dangerous winter herring season that, from the 1930s to the 1960s, ran from the end of salmon season in November through to early March. This was in the days when herring were sold to the reduction plants for processing into meal and oil. Generally, the herring fishing would be in the Strait of Georgia before Christmas and "up North" after the holiday.

Richard's wife Francis says, "Our wedding anniversary is on December 20 and Richard was always home before that to put up the tree with the lights and then go to bed. It was kind of close sometimes but he always made it. Some of the boats had Christmas trees on the mast when they came in."

ALWAYS HOME FOR CHRISTMAS

Christmas shopping in Ladner was dependent on the salmon cheque and Francis remembers doing her shopping from the Eaton's catalogue. "In those days it was a real trip to town. You went in the dark in the morning and came home in the dark at night." Of course, this was in the days a ferry transported passengers across the south arm of the Fraser River, before the tunnel was built at Deas Island in the 1950s. In those days, Ladner had much the same isolation of an up-coast fishing community and people made their own social life. "This bunch was an awful partying bunch," Francis remembers.

"We would start from the time we came off the herring and go from house to house," Richard adds.

And then there were the dances. The community hall had the New Year's dance, but the Ladner Fisherman's Ball was the biggest dance; it was started close to fifty years ago and was held early in December. It hasn't been held for years, perhaps because people got more involved in different kinds of fishing that demanded different schedules, or perhaps times have just changed. Ladner is now at least partly a suburb of Vancouver, with all of the city night clubs a half hour drive north through the tunnel.

In spite of the condominiums and fancy float houses, Ladner is still very much a fishing community. During December the docks are crowded with boats with empty net drums while their skippers and crews visit over dining room tables and catch and re-catch all of the fish that swim and have swum in the sea. In Richard Martinolich's house, the children and grandchildren gather until twenty or more people have joined to celebrate. They will reaffirm the traditions that came from the Adriatic coast to Ladner, and that are still a part of these Canadian fishermen's Christmas. 🐟

The *BC Kid* (above), shown here in 1940, was one of the first seine boats built in BC. Photo: Lloyd Fuerst.
She was sold by the Martinoliches, renamed *Homelite* and was still fishing in 1996 (below). Photo: Alan Haig-Brown.

Nimpkish River

THE ELUSIVE NIMPKISH DOGS

Don Pepper

Salmon seine fishermen know that each species of salmon heads for the spawning grounds in its own way, and you have to know this to catch them. When fishing pinks one can tow the net forever. They are slow moving, somewhat docile and will not back out of the net. Sockeye, on the other hand, are fast moving and will back out of the seine if you tow too long. But to my mind, the trickiest and most difficult fish to catch were Nimpkish chums or "dog" salmon (alas, they are no more). I had the opportunity to fish them in the 1950s when I lived in Alert Bay, across from the Nimpkish River. In the late fall we would fish Nimpkish dogs and many a story was told of their wily manoeuvres. They were said to be impossible to catch, but of course we did catch them—sometimes.

In the late 1950s, before the widespread use of net drums, we still used the Puretic power block to haul the seine net aboard, and the seines were made of cotton, tarred, and sometimes "cutched"—which was the application of some sort of bark extract as a preservative. We had Spanish corks (real cork from trees) to float the net, and the lead lines were three-strand manila with leads strung on them like beads. Not your fancy braided nylon with the lead embedded in it. Our purse lines were four-strand manila which had to be stretched and coiled just so. The secret was to get the kinks out of it so it would coil easily. Splicing in the connecting links for the "figure eight" was a backsplice with a taper. The figure eight was a brass connector that joined the purse lines together. Seine gear was much different than it is now and the pace of fishing was definitely slower.

In November 1959 I fished for Nimpkish dogs with "Boots" Jolliffe on the *Moresby 3*, a nice little seiner based in Alert Bay. My brother Ron was really the crew member, but he had a job in the local bakery so I substituted for him on a two-day opening in early November. My motivation was money, and the slight chance that Boots may at some time take me on for the halibut fishery. That never came to pass, but I had hopes.

The fishery opened on Sunday night at six, as was the custom then, but as it was late fall and dark at that time, we didn't bother to go out. At five o'clock Monday morning I woke up at home, dressed, and met Johnny Lawson at his house a few doors along. We walked from Pepper's Point (where I lived, being a Pepper), which was the end of the road in Alert Bay in those days. As we went along the road from my house to the Shell Oil dock, the crew assembled. First we picked up Boots, the owner of the *Moresby 3*, then Ed "Gopher" Gordon, and finally Jake Smith who was already at the boat. Soon, Johnny had coffee going, and we were underway by six o'clock.

THE ELUSIVE NIMPKISH DOGS

Illustration by Graham Wragg

At daylight we made our first set at Blinkhorn. I tied to the same stump I had tied to a million times. I examined it last year from the bridge of the *Prosperity* and it is still there. The tide was the first of the flood and we were to follow the Blinkhorn/Double Bay cycle of sets known to all Alert Bay fishermen. We didn't get much and we tried a few other sets but no luck. Then we went to Bauza Cove. This was a favourite haunt of Nimpkish dogs, as the flood would shove them out of Weynton Pass across Johnstone Strait into the cove. They were not heading south like sockeye, but heading back north to get to the Nimpkish River, seven miles away. To our simple minds they were going the wrong way.

What made Nimpkish dogs so tough to fish? Was it because we were dumb? Or our gear inadequate? No, experience told us they were difficult to find because they would school up in strange places, Mitchell Bay or Rough Bay in Sointula, Carr's Place ("the Pig Ranch"), Parson's Bay behind Alert Bay, or stick around Haddington Reef. They would show sometimes by jumping, but that gave no clue as to which way they may go. Then they would fin, an inch or two of their backbone showing, and your eyes could play tricks on you. Was that a school of fish or a wavelet or merely a reflection in the water? When you did set, they may go the other way and never enter your net. Or you would see them in the net, but at the stern of the boat going out. Either you got them or you didn't. An empty net was called a "skunk." I truly believe that more skunks were pulled on Nimpkish dogs than any other fish.

These thoughts were on our minds as Boots took us into Bauza Cove. This was unusual, as this was not the normal setting spot, but Boots had seen a few finners and concluded that the southeast wind had blown the fish into Bauza Cove. The dogs were schooled up there and not moving, biding their time until they went to the river. We made the set, nice and slow to let the lead line sink, came around, pursed up and brailed aboard 500 big Nimpkish dogs.

The set sticks in my mind after all these years because we made more sets that day and the next, but none as simple and successful as that one. I remember the big sets we did or didn't get since that time, but it is the neat little unusual sets that I really like. Boots was a master at finding fish in strange places when no one else was getting any. And so it was that day. We went home and I slept in my bed and was up again in the morning to walk the road to the boat with the southeast wind at our backs. We tried two sets at Blinkhorn for nothing, then to Haddington Reefs, and then we tied to the dock in Sointula and watched for jumpers. Nothing showed so we went home early. We delivered our fish to the packer in Alert Bay. When we were done, we stripped the net, cut off the lead and corkline hangings, put the web into tanks full of bluestone, a copper sulphate preservative, hung up the purse lines, put the brailers, poles, plungers, and lines in the net locker and went home. I was home in time for supper. A simpler time, a simpler fishery, but fun.

PUTTING SOMETHING BACK
Billy Griffith of Egmont

Roxanne Gregory

Egmont

"...enhancement is the way we honour the salmon." Billy and Iris Griffith. Photo: *Westcoast Fisherman* collection.

Nestled at the north end of the picturesque Sunshine Coast, just south of Earls Cove, is a narrow, winding, black-topped road leading to the tiny community of Egmont. By navigating the occasionally steep switchback and travelling through a warren of back roads surrounded by mountains and the sea, you finally reach Billy Griffith's driveway near the end of Maple Road.

On the left of the driveway is a boat ways with an enormous seiner erupting from a skeletal framework of wood. A tall, lean man with wispy red hair and weathered eyes is busy fitting planks in her hull while a younger man helps. Billy Griffith is a man of the sea, one of the old-timers who first worked on seiners when nets were hand-pulled. He has seen a lot of changes on the coast and in the fishing industry—not all of them for the better.

The modest home he shares with his wife Iris is not far from the ship he and his son John are rebuilding. Between the ways and the house is a large building that serves as a private museum for Billy's collection of marine engines. Easthopes and Palmers and Vivians, some working some not, crowd the floor and surround the walls, smelling of grease, gasoline, and diesel.

Billy grew up in Egmont. His grandfather homesteaded a piece of land near Skookumchuck Narrows in 1920, and Billy's father raised his family here. Billy and Iris raised their family in the same house. Billy remembers the early days of World War Two: "When I was a kid, most everybody fishing owed the Japanese store owners, including my dad. I remember when they took them away for internment during the war. Afterwards, when they

came back to BC, my dad went to Vancouver and found them. He paid them what he owed them, and they told him he was the only person who did so. He used to visit them every year when he was finished with business in town."

Billy's commercial fishing memories span almost fifty years. "I first went crewman with Bill Silvey on the *Dollar II*, in 1949. That was in the old days, when all of our local boats were pulling net in by hand. There were no drums or power blocks. We had a live roller on one side of the boat, driven by chain, and a dead roller—that is to say, a free-wheeling roller—on the other side. If you pulled hard, it would turn, if you set on that side of the boat. So, usually, we only made sets on the side with the live-roller. It was hard work in those days. Even though the nets weren't very big, they were made of cotton web, Spanish corks and heavy gear...the skipper pulled lead line, and I was just a kid and I couldn't pull my share of the web. I was ashamed of that, but another crewman said, 'Just pull what you can, and we'll do all right.' And we did. It was good to work for a guy like that, especially for a young guy just starting out."

Billy's fascination with marine engines began in 1950, when he went out with Leonard Silvey in the 29-foot *Leonard S*, a seiner with a 5-horse Easthope. "It was a pretty small boat, pretty haywire, with a winch that worked okay until we got to the heavy part. And then the pin in the winch drive would shear off and we'd have to finish pursing [the net] by hand. The skipper was the only one who could start the engine. I was used to Easthopes, but that one I couldn't start, it was so badly worn out.

"In those days, the steam boat inspectors wouldn't even condemn boats like that, they just shook their heads and walked away. That's all you had then. It was a pull-yourself-up-by-your-bootstraps show. And sometimes they weren't very good bootstraps you were pulling on. Nobody had any money. You never went hungry, but there was no money for anything, either. The tiny bit of money you did get, you had to spend on equipment, to try to patch yourself up for another day, so you could fish, so you could pay for the fuel and the parts to patch yourself up again. It was still pretty hand-to-mouth.

"I'd log in the winter to make my payments, and quit in the spring. I guess I wasn't too good a fisherman then. I had a bunch of close shaves in the bush in '57 and I thought my luck was going to run out if I kept pushing it, so I never went back to the bush after that.

"I gillnetted by hand, but that was no fun. Drum seiners and power blocks came in during the mid-fifties. A lot of the early ones had problems with the drives. The idea was good, but they hadn't learned how to hang the nets. The roll-ups and backlashes and mechanical drives they had rigged up weren't strong enough, and things would break. When they went to hydraulics, they weren't strong enough, either. Pipes would burst, and seals in the motors would go. It was just heartbreaking. I used to watch them drifting around, and we'd laugh at those drum seiners. Now we are all drum seiners."

Billy and his father built the *Tegula*, a 35-foot seiner, which Billy ran in 1955. Although it was a small boat, he took the trouble of fitting it with

Billy among his eclectic collection of engines and gear from yesteryear. Photos: Roxanne Gregory.

a revolving platform, called a table, making it one of the smallest table seiners on the coast.

Shaking his head, Billy says, "Fishing has changed a lot since we first started. I used to fish five days a week from June 'til November, all up and down the coast—very few closed areas. It used to almost all be open. Mostly, you could fish when you wanted and wherever you wanted. If we kept that up there would be nothing left to fish. If we hadn't started conservation measures twenty, thirty years ago we would be facing what the loggers are facing today. Look at what has happened in the Atlantic to cod stocks. On the coast, the trollers are probably the worst off for regulations now. But, if we didn't submit to these conservation measures there wouldn't be a resource left, no industry at all. But it does make it hard going."

Billy thinks the federal Department of Fisheries and Oceans does a "fair job of resource management," but he is worried about the future. "It's a difficult job," he sympathizes. "You can't see the fish, and you have to manage a resource you hope will be there. Slowly we're building up some of the resource in these closed areas. There are 400,000 sport fishermen in the gulf every year, and that has had a big impact on the spring salmon."

Active in salmonid enhancement for the past fifteen years with the Pender Harbour Wildlife Society, Billy talks about restoring the resource: "In Victoria, a group of trollers starting working at Goldstream River to enhance the dog salmon stocks. I think they were the first commercial fishermen to get involved. I got involved because I felt we had taken a lot out of the resource and I thought we should put something back in. Like the natives who honour the salmon in their religion, enhancement is the way we honour the salmon."

Fishing is an ongoing family affair for Billy and Iris, son John and daughter Maureen. He is positive about the future: "I think we'll always have fishing as an industry, but we have to work at it, and we have to be conservation minded. We have to put something into it, if we want the resource to be here tomorrow."

The Griffith wharf at Egmont. Billy's father homesteaded this piece of land near Skookumchuck Narrows in 1920. Photo: P.M. Foss.

Johnstone Strait

Family and crew of the *Western Investor* (left) and *Corregidor* in 1987, waiting for the Johnstone Strait seine fishery to open at 1800 hours on a Sunday. In the foreground the next generation of Robertses is introduced to fishing.

ANOTHER GENERATION FISHES THE TIDES
The Roberts Clan

story and photos by Alan Haig-Brown

Aubrey Roberts and his brothers are well known on the BC coast for their skill in seining the tides of lower Johnstone Strait. But now there is a new generation of sons and nephews taking their turns at the tie-up places between Harry Moon's (Camp) Point and Chatham Point. The favourite spots are those with the trickiest tides, like "the Slide" and Humpback Bay, both up above Bear River. Below Windy Point, the tides are a little more forgiving, but still no place for the uninitiated.

The Campbell River, Cape Mudge, and Salmon River people have been fishing these waters since before recorded history. When travel was by canoe, back-eddies would help a tired paddler gain the next point. These

ANOTHER GENERATION FISHES THE TIDES

same back-eddies collected up the schooling salmon, just as they do today in places like "the Pineapple," where those who know how make a half-set in the back-eddy on the ebb tide.

Aubrey Roberts got his start in fishing back in the fifties. By 1958 he was a seasoned crewman, although still in high school. The town may not have understood the complexities of the tides, but they understood when Aubrey and his brother Gerry drove to school in a Jaguar sports car that they must be doing something pretty special. Aubrey left school to pursue success on the fishing grounds of Johnstone Strait, and eventually to fish deep-sea tuna. But his son Bradley, born in 1964, took his dad's advice and finished high school, although he couldn't help noticing that it probably cost him $100,000 in lost earnings from fall fishing and herring.

Brad's mom, Jenny, remembers that she stayed home from fishing the summer that he was born, but the next summer she was back out with her just-under-a-year-old baby. Brad remembers that he tried staying home from fishing a few times when he was a teenager, but he always rejoined the boat before the season ended. "I tried a few other jobs and nothing seemed to come quite as easy as fishing always has...I was going to try being a longshoreman with a buddy of mine; but every time you start a new job, you start right at the bottom of the barrel. I think what I really didn't like was not being on the water."

When he was nineteen, Brad started to hear from his family that it was time to start thinking about skippering a seine boat. Then, when he had a car accident and missed herring season and was not in good shape to work on deck for salmon, the decision was made. He became skipper of the *Anna M* in the summer he turned twenty. Brad explains that the first year fishing the tides of Johnstone Strait, you "kind of wing it. I had a pretty good idea what was going to happen. But there are a lot of finer points to it. Older guys like my Uncle Tony or Dad will see the tide move away from the beach and they'll try a quick one. I was never quite that brave. I knew that I had to get a big lead if it was a big flood or watch to get pushed into the beach and keep a good bag in the net to drum off the beach and all that kind of stuff. Basically, what everybody else knew. You just kind of have to sit back and watch."

Brad would ask his dad for advice on when to make a set. "It pretty well got to be a habit after a while. He wouldn't let me make a mistake, but if I started getting too simple with my questions he just wouldn't answer me."

In 1986, after two years of learning the tides and a lot of finer points of skippering with the *Anna M*, Bradley decided to buy the *Corregidor*. "Ever since I was a kid, any time I thought of a nice wooden boat it was

Bradley Roberts at the wheel of the *Corregidor*.

Maxie Chickite stands by to start the set by knocking the pin out of the skiff's quick-release mechanism.

Simon Dick and Cory Wilcox "standing by" for a drift set.

The crew of the *Corregidor* preparing the net for a second set in the evening light.

always this one that came to mind. When somebody told me that it was for sale, I just wanted to leap right on it."

The *Corregidor* was built in Tacoma, Washington in 1942 for the pilchard fishery, but with the decline of that fishery she and a number of other boats such as the *John Todd*, *Mary Todd*, *Western Ocean*, and the *Key West* were sold to Canadian fishermen. Even for a big seine boat she has unusually heavy construction, with 3-inch planking and lining on 8-inch sawn frames for a total of 14 inches of hull thickness. She is twice the age of her skipper, but Bradley sees a long and productive life ahead of her.

While Brad's dad had encouraged him to go to college, in case he wanted to leave the industry, Brad had no such intentions. "Unlike a lot of people, I've got a lot of faith in the fishing industry for years to come." 🐟

Illustration by Alistair Anderson

MY BUNK

Michael Skog

Afisherman's bunk is his castle. Can anyone who has spent a significant amount of time at sea dispute this? I suggest that this
berth is much more than just a bed. It's a place of peace, security
and rest, the only place on a large vessel where one might find repose. To a
crewman, a bunk has a womblike quality. It becomes an illusion of privacy
on a boat sometimes overrun with rustic companions. Occasionally, it
becomes a refuge from insanity. But what happens when something goes
wrong with one's sacred spot?

My bunk was a haven of solitude. Generally speaking, the crew got
along well, and the skipper was even a great guy. There were five of us seining salmon that year, and most of us were leftovers from the year before.
Yet, there was always a time when we each needed our own space, especially from two crew members whose idiosyncrasies laid upon the psychot-
ic edge. The place where I could escape was my bunk—the prized bottom
bunk. It was my hermitage. It was there where I closed off the outside world
with the drawing-together of the short blue plaid curtains.

However, throughout my last year aboard that vessel, I discovered that my Shangri-la was violated by a curse. This was revealed through a series of episodes that began with a shipmate—a man whom I protect with an alias. (I shall refer to him as "Petri Dish" after the laboratory device used in growing bacterial and viral cultures.)

One morning, after a salmon delivery to Vancouver, I returned to the boat to find the Petri Dish sleeping in my bunk. His naked butt faced insolently upwards. I was tempted to pull him out and toss him to the floor, but instead, I chose to wait until he sobered up. If there was a reasonable explanation, then perhaps I could understand the liberties he had taken with my personal sanctuary. As time passed, I began to doubt the existence of an excuse adequate to calm me down. The rest of the crew was expected to arrive shortly, so I made some coffee. When they got there, all of them went in to have a look at Petri's pink, hairless rear end. The bunk hustler awoke to the sound of laughter and soon emerged into the galley followed by a large swarthy maiden. This woman did not look like she was a stranger to strangers and I became horrified at the thought of the two of them romping on my sheets. Everyone was joking with the Petri Dish about his current conquest, so I thought it would be inappropriate for me to hit him, or even give him a piece of my mind. After her departure, however, I inquired into the exact location of their nocturnal encounter. Sensing that I might be annoyed, the Petri Dish assured me that it was not in my berth but his where all the action took place. His was the bunk above mine, which required a bit of agility to reach. I questioned the likelihood of two gassed-up, middle-aged swingers risking broken necks in order to copulate against the ceiling. My suspicions were later confirmed when after tending to the net, I discovered lipstick on my pillow. There was nothing left to do but wash the sheets with gallons of bleach and pray that I would not discover a rash which should belong to someone else.

Another of these episodes occurred when the skipper's son joined us for a trip and was temporarily placed in the berth directly above mine. While travelling along the western shore of Vancouver Island, somewhere off San Juan, I finished my wheel watch. I crawled into my bunk, but not before noting that the boy was having a restless night. I fell instantly asleep to the gentle swaying of the groundswell. After two hours of slumbering deeply I was partially awakened by a vision of my skipper wiping the curtains and the outer walls of my bunk with paper towels. It was such an uncommon sight that I determined it was a dream. I went back to sleep.

The next morning I awoke to the acrid smell of bile. As my head lay upon my pillow with my top eye open, I began to make out colours on the linen where the night before there had been none. I lifted my head and looked closely. I saw my pillow and surrounding sheets splattered with chunks of partially digested food and yellow rivulets of bile. It was around this time that I first began to suspect a curse had been placed on my bunk. A few weeks later, I was convinced of it.

On the seiner was another character—the Old Man—whom I will also protect with an alias. He was a man whose world was shrinking. A product of last century's social codes and alienated because of a hopeless

MY BUNK

grasp of English, he was a threatened man who bragged that women should be controlled with a two-by-four. Hopelessly lacking a diversity of skills, the Old Man could not survive anywhere on the planet now, except hidden away in the engine room of the old seiner. One weekend he, the skipper, and the Petri Dish decided to stay in Port Alberni and babysit the boat while the cook and myself bussed back to our homes. After this short vacation, I came back in time for the following opening. The fishing went pretty well, although I could not shake a clammy feeling I had the entire trip. There was also an uncomfortable aroma, which I attributed to the boy's mess a few weeks earlier. After we delivered our fish and tended to the maintenance of our net, the Petri Dish discreetly mentioned that I might want to wash out my sheets. After I asked why I should do this, he reported an adventure he had with the old man while we were all away.

Apparently, the old man got into a jug of his homemade wine, had one hell of a time, and then passed out. Upon drinking so much fluid, however, he woke up with a need to relieve himself. Unfortunately, he took a wrong turn, and instead of finding himself at the railing he found himself at my bunk. There he found bliss for his strained bladder.

"You're telling me this now?" I asked, unable to contain my horror.

"Well, the Old Man made me promise not to tell anyone," said the Petri Dish.

I barged into the crew's quarters and stripped my sheets and hauled out my abused mattress. In a show of frustration I threw it on to the dock. At that moment I wanted a handgun to do some damage to the Old Man when I found him. The bunk to me was like Bikini Island after the atom bombs had fallen. How could I ever enjoy that refuge of solitude after all the indignities it—we—had suffered. I finished off the season, but never returned. Good riddance! Perhaps I could have forgotten the whole thing if it had been my car, or leather jacket that was compromised. And maybe I could be laughing about it now, years later. But, damn it! That was my bunk! 🐟

Strait of Juan de Fuca

FACING OFF ON THE BLUE LINE

Don Pepper

When Pierre Trudeau's father-in-law, Jimmy Sinclair, was federal Minister of Fisheries, he started a new salmon fishery in the Strait of Juan de Fuca. He did it to put pressure on the Americans to sign a Fraser River pink salmon treaty. To encourage BC fishermen to fish there, he made it easy for them to acquire large boats from the collapsed sardine fishery in California. This was in the 1950s. I fished there in the 1960s on the seiner *Barkley Sound* with Freddy Jolliffe to get money for university. I said I would never go back, but I did in 1994 on the *Prosperity* with Byron Wright (who was also there with me long ago). I was there in 1994 as part of the 1950s reason: to put pressure on the Americans to sign a treaty.

Fishing Area 20-1 is different from any other area of the coast. First of all, seiners are allowed power skiffs and larger nets than on the "inside" coast. Second, the boundaries are special. The outer boundary is the US–Canada border; the inner boundary is the 30-fathom ribbon boundary (to protect spring salmon) along the Vancouver Island shore. But the crucial boundary is the one that runs from Bonilla Point (below Carmanah Point) to Ta'toosh Island (on the US side). This is the so-called "blue line." It got its name after the line in hockey over which you are not allowed to go without the puck. The commercial fishing blue line limits where you are allowed to fish. It marks the first opportunity for the seine fleet to intercept Fraser River-bound sockeye and pink salmon returning down the west coast of Vancouver Island from the high seas. If you go over the line, fisheries officers make you dump your fish. This is complicated by the fact that the tide generally ebbs to the blue line, so you are always in danger of arrest.

Because it is a tricky spot, and many boats are crammed into the box created by the boundaries, a certain protocol has developed about how you fish there. It all revolves around a "starter boat." The starter boat is whoever gets there first for an opening. He selects the depth he wishes to fish in. The next boat arriving asks the starter boat for a "turn" and calls his preferred depth. The third boat asks the second and so on. The crucial point here is that you cannot get a turn unless it is given to you by the legitimate holder of the last turn (more on this later). Each boat expects a quarter mile of distance between him and the others. Naturally, the quarter mile range ring on the radar helps. When all the "good" spots on the first line are filled up, a second line forms with its starter boat.

On an opening (usually 0700 hours), there can be a series of lines all the way to Port San Juan, 10 miles away. The distance between the lines is not fixed, but a 20-minute tow on a 1-knot ebb tide means you could

FACING OFF ON THE BLUE LINE

cover about a third of a mile of "fishy water." That is a minimum; a mile is common. Distances between the lines vary as to circumstance. But once fishing starts, the boats on the blue line effectively block the fish to the other lines. The result is that the lines soon collapse and boats queue up in the various depths on the blue line.

It is a nightmare. With up to two hundred seiners working in the area, it is difficult to get clear radio time and to find out who's who and what's going on. It starts out simple and ends up confusing.

Here is how it is usually done. The fleet has anchored up close to shore by Bonilla Point. The "starter boat" pulls his anchor about 0630

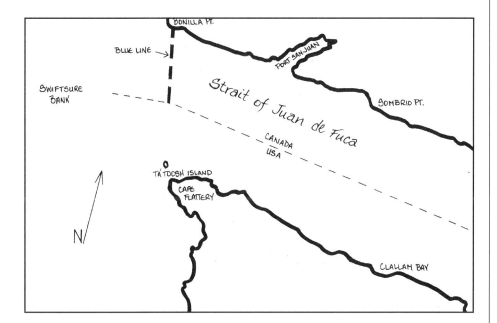

hours and runs out to his depth and the fleet follows him out. If it is an ebb tide, the starter boat heads to the east, away from the blue line, with every-one following a quarter of a mile from the boats inside and outside of them. As the first line moves down, the second line wants "fishable water," so it moves to the east also. At the opening, each boat sets his net and if it is done correctly, the bow of each boat is next to the skiff of the outside boat and there is a solid line of nets on the blue line and each line behind it.

It is after the first set that the lines collapse and boats then call a turn in the various depths on the blue line and some have already called for the second set. That is, there is already a queue on the blue line, as some would rather be the second set on the first line than first on the fourth or fifth line. Obviously, some depths are more favoured over others and the line-ups in these are long. Some prefer to be "in the deep" and fish near the Canada–US border, others wait for the flood tide and move to the shallows. Boats are continuously moving from line-up to line-up, but one constant is that there is a net in the water at all times on the blue line.

From all the confusion, certain unwritten rules have emerged. The cardinal rule is that you cannot get a turn in a depth unless the legitimate holder of the last turn gives it to you. Then you can give it to the next boat. You can call for a turn in any depth but you must be there to collect it. A boat that gets there before you can pre-empt your turn, and this does produce some arguments. Further, the process of getting a turn is not as neat as it should be. Typically, a conversation goes like this:

"Pete, you last after Wally in Sid's line?"

"No. Mike is."

"Mike, after you. Okay?"

"Okay."

With twenty other conversations going on, nobody but the two know what the hell is going on. I know, I've tried to keep the line-ups. Keeping track of six line-ups with ten boats in each and people changing their minds and switching lines and with other conversations going on, it is radio hell. There is, I guess, a correct procedure. It might go like this:

"*Lulu Island*, are you last in the 55-fathom line-up?"

"*Lulu Island* back to the boat calling. Yes, we are last in 55."

"Fine, I'll take the turn. *Nishga* after the *Lulu Island* in 55 fathoms."

But that would be too simple and clear. Not a procedure for the happy confusion of fishing the blue line.

By the way, the fishing was lousy (thanks to El Niño) and the wall of nets we put out there in the 1994 fish war caught few sockeye. But should you ever have to go to the blue line to fish, you now know how to call a turn. It will be confusing and the amount of fish you catch will, as usual, be uncertain. Some things are confusing and uncertain. I wonder if Jimmy Sinclair mentioned all this to Pierre Trudeau when Pierre asked for his daughter Margaret's hand in marriage? 🐟

Illustration by Alistair Anderson

PRANKS FOR THE MEMORIES

Michael Skog

I liked being a fisherman. I especially liked salmon seining. I got to travel the same beautiful routes as the cruise ships. Ate like a king. Had time to hike around shore and fish for sport. It was like being paid to camp. It sometimes had its drawbacks, like when fishing got bad, but one thing it did more than anything else was extend my childhood well beyond elementary school.

Take one summer, for instance. I signed on as beach-man aboard the *Elling K* in the last half of June, and by early July, I was scrambling along the shores of the central coast. This season didn't start off very fulfilling. In the first week the fishing was bad, and it stayed that way for the rest of the month; the expenses were forcing us to dig deep into our pockets. The whole crew's spirits were down. Yet the worst of it, for me, was being away from big, or small, city stimulation—that whole month we delivered only to packers.

Isolation was difficult. We were the same five guys, stuck in the

same boat, no VCR, trying really hard not to repeat the same stories over and over again. Days passed slowly and our patience wore thin. We needed a change. Women—we missed our girlfriends and wives. We also missed the soothing presence of females as a whole. Their surrogates, found in the girlie magazines, only taunted us with their pretence of availability. Hormones began to flow through our bodies, making us extremely agitated. Not since we were all seventeen had our minds been so preoccupied as a result of those infernal books. We started snapping at each other for the smallest offences. Things were beginning to get dangerous. Times would be better, we thought, if the fishing would improve.

We would cling hopefully to rumours about a possible west coast sockeye salmon opening only to have these hopes mercilessly dashed when fisheries officials postponed it until further notice. Finally, relief came in the form of a rare opening of the Inside Straits for pink salmon. Finally, a change of scenery, but where the hell were the Inside Straits, we wondered? They sounded more like a poker hand than a fishing ground. After we returned from there, we were able to deliver to Vancouver 300 of the scrawniest fish I ever saw.

The one night we spent in town completely rejuvenated the crew. On our way to San Juan the next day—its opening had been announced while we fished pinks—the crew of tough, brooding juveniles I shipped with in the Central Coast had transformed into giddy kindergarteners. I thought I was on the wrong ship.

That evening, during my wheel watch, I was reminded of the light composition that was the fisherman's soul. During my four-hour spells at the wheel, I always went down to the galley to drink a glass of water from a jug I kept in the fridge. This ritual helped to keep me alert during the night. I still do not know who, but someone spiked my water with a large amount of dish soap on that evening. Honestly, while I coughed half to death, bubbles shot out of my mouth and nostrils—just like on the cartoons. I had to admit, that was a good one.

The next day, Rick the cook got Mike the skiffman. While Mike was reposing in his bunk, Rick filled the sleeper's hand with a mound of shaving cream—an old joke that will always be a classic. The cook then tickled the sleeper on the nose. I was not there to see what happened, but I did see Rick flying out of the sleeping quarters laughing hysterically and holding his gut. Mike emerged seconds later with foam streaks all over his face and shirt. This was a popular prank, which never completely ran its course. On two different occasions, I awoke just in time to catch the silhouette of a suspicious-looking crew member standing over me with a can of shaving cream in his hand. Revenge or mischief was on everyone's mind; life was again becoming very dangerous on the boat.

Our cook was an industrious fellow. He was determined to catch himself a salmon before the opening. While we were anchored, still awaiting the opening, the tide flowed swiftly under us. Rick hung a trolling rod over the side of the boat while he was cleaning the galley. As he was busy washing the dishes, he didn't notice that his rod caught a mud shark.

Without telling Rick, Mike reeled in the squirmy fish and set out the line again, only this time he brought a bite of line around to the other side of the boat where he pulled on it, vigorously. He called out to Rick, who nearly fell on his face trying to exit the galley door. While the cook reeled like a man possessed, Mike slipped into the galley through the other door and placed the live shark into the dishwater. As Rick returned, he was telling a tale of an immense battle with a colossal fish. Anchored this close to shore, he had to have caught one of those big spring salmon. It must have been foul hooked too, for it got away in an enormous burst of speed just as it neared the boat. He began to relay a similar incident, which had happened not too many years before, as his hands lowered into the sink. For a microsecond, I saw his body assume a confused posture, as if he were trying to make sense of the sandpaper-like flesh that squirmed in his hands.

I have never known a man to shriek like that before—like a cat with its foot trod upon. His arms flew above his head and collided with explosive force against the low galley ceiling. Simultaneously, he sprang away from the sink and nearly rolled across the table. I don't think it's right to laugh at other people's pain. So, I'm sorry to confess that I came very close to soaking my pants that afternoon—as did the rest of the crew sitting around the galley table. Tears rolled down our faces as we clutched our aching bellies. Even Rick, after he recovered from his initial shock, thought this was a pretty good one.

Was this a typical way for grown men to behave? I don't know. I don't know because I don't think I know any grown men—the only men older than myself all happen to be fishermen. That's the real reason I always liked fishing; there's no pressure to become an adult.

Say, that reminds me, I still owe someone for spiking my water jug.

Seymour Narrows

This photo taken from a Coast Guard rescue helicopter shows the upturned hull of the *Miss Joye*, in which Maxine Matilpi was trapped for two and a half hours. Photo: Maxine Matilpi collection.

MAXINE MATILPI
Capsized in Seymour Narrows

as told to Vickie Jensen

I've been fishing for most of my life. My grandfather Henry Speck was a fisherman. My father, Charlie Matilpi Sr., was a fisherman. All my brothers fished. My first husband was a fisherman. He operated the seine boat *Star Wonder*.

The accident happened when we were out on James Walkus's boat *Miss Joye*. It was my first season for herring, and it was probably the third week we were out, near the end of November. There were seven adults on the boat: Kenny Lambert and his father Forest, Bruce Rafuse, Gary McGill, Fred Anderson (my ex), myself and Kenny's wife Betty, with her six-month-old baby, Jason.

It was rough that morning, and we had travelled all night from Ganges because James wanted us to go to Deep Water Bay. He'd said that

MAXINE MATILPI

the first boat to catch 35 tons would get to go home after delivering to Vancouver and would get the weekend off. *Miss Joye* was the first one to catch the tons, but instead of us going home, he pumped the herring from our boat to his boat and travelled south to deliver while we travelled in the opposite direction.

We were supposed to stop in at Campbell River for at least a couple hours for all the crew members to get proper rest, to get groceries and fuel up. Betty and I and the baby were in the galley, and we thought for sure we were going to pull in to Campbell River. But as we got closer we looked out the window and Betty said, "Holy shit! We're going through instead of stopping." The tide was running at its full ebb through Seymour Narrows, but we kept on going.

It seemed like not even five, ten minutes into the Narrows when we hit the first whirlpool. It was ever so slow. We didn't know what we had hit then, but when we came creeping back to position and travelled further along, we could see it. We had rolled way over, so that if the window had been open, we could have touched water. We just braced ourselves. Betty said, "Oh my God, I hope nothing happens to me." I said, "Betty, don't talk like that." For some reason she felt fear that day, and she was not a person who was easily scared.

Kenny was the captain. As we were going through, he was on the radio trying to talk to James to tell him that we were going through rather than stopping to wait for the tide to slacken off. We were trying to hug the shore. Bruce was on the wheel. It seemed like it took all his effort to control the boat. I remember looking into the pilot house, and Kenny had his legs braced far apart. Normally, a lot of boats will pull back the throttle, shut the engine off and just glide through. But for some reason that morning while Bruce was on the wheel, the throttle wasn't pulled back until we almost were into the second whirlpool. Fred, my ex, had just gone to bed along with the other crew member, Gary. Forest was awake; he was the engineer. We'd had engine problems that night, so he kept getting up to check on it.

That boat was so cranky, so top heavy, that anything could've happened even before we got to Seymour Narrows. Plus we didn't put the net in the hatch to stabilize the boat. Then we hit the second whirlpool. The first one that got thrown was the baby, who was sitting in between us on a table in a cuddle seat. He went flying and hit his forehead on a counter by the sink. Betty and I got thrown, too. She was the one that was closest to the stove. I remember there was a huge pot of boiling water on, plus there was a big kettle that was almost to the boiling point because we were planning to make stew for dinner that day. We were sprawled out on the galley floor, which was at an angle. She kept on saying, "Keep his head out of the water. Push him up." But by the time we got hold of Jason, he was already limp. I knew right then and there he was dead.

You could feel the boat tipping over more and more, then the water started gushing through. It seemed like it came from the pilot house first. I just remember feeling the cold, cold water. Betty was by the stove, and the

For Maxine, her two children Aubrey and Mandy provided a reason to keep on living after the accident. Photo: Maxine Matilpi collection.

big pot dumped all over the right side of her body. All she kept on saying was, "Oh my God, oh my God. Keep his head out of the water."

I could hear Kenny trying to talk to James, but the phone was breaking up terribly. By then the throttle got pulled back. When the engine shuts off, a bell rings in the engine room. My ex heard it and came running up in his underwear, saying, "What the hell is going on here?" We were ankle deep in water. Betty picked up the baby, and Kenny came right behind Fred and said, "Every man for himself. Get out of the boat. Get out of the boat. We're going to sink."

Kenny and Betty and the baby went towards the pilot house to get out that way. Fred and I were in the galley, trying to open the galley door, but there was too much water pressure. So we grabbed a frying pan and kept banging the window by the sink, but it just kept bouncing off. By then we were almost standing on the stove. The water was coming up fast, really fast. We were three-quarters upside-down by then. By the time it got to our chests, we were completely capsized. I don't know how many times we tried to get out of that window, and we just couldn't. Finally, we just hung onto the lazy susan with our noses pressed right up against the floor to get any air.

I don't think it took that boat fifteen minutes to completely capsize. Meanwhile, Bruce, Gary and Forest got out of the boat. The net had already started to unwind from the drum. Bruce had a pocketknife on him so he was able to cut himself out of the net. That saved his life.

Forest got free of the boat but was too exhausted to swim back to it. He hung on to this wooden refrigerator box, but the current was too strong. The last they saw him he was hanging on to this box. When they turned around, he was gone. He never surfaced again.

In the galley, I could feel somebody tugging at my legs. I thought this person who was tugging on me wanted to breathe, so I kinda kicked my feet to let them know there was air up here. But I was to afraid to go under the water. That happened twice, and then the person just quit. I thought it was Betty, maybe, but it was actually my ex. He had dove under and finally kicked the window out after we ran out of air. The window was quarter-inch-thick plexiglass; he didn't realize that he had cut his foot badly, all the way around. He cut a main artery. When he got up to the surface, the crew members helped him to get to the keel; that's when he realized that he was bleeding. The cold water saved him from bleeding to death.

All three guys were on top of the bottom of the boat by then. Underwater, I found my way to the pilot house. The interesting part was when I decided to leave the galley, it seemed like I felt my grandfather's presence, Henry Speck. For some reason I just had to follow this feeling that told me to get to some light. So I dove down and saw this light like a lantern, and I followed it. I came up to surface, ready to burst, and sure enough I had at least five feet of air in the pilot house. I stood on the wheel, but water started coming up really fast. That's when I started to panic and screamed, "Please God. Please God. Don't let me die." I thought of my uncle who had drowned years ago and what he must have gone through.

MAXINE MATILPI

When water got past my nose, I thought, "Now I'm going to die. This boat's going to sink, and they're never going to find me." So I dove down again. Again I followed this feeling that my grandfather was there guiding me. This time I swam towards the engine room. It was pitch black down there. There's a kind of wraparound stairway, and as soon as I got to it I knew exactly where I was. I just collapsed. I had no energy. I couldn't even lift my legs up.

There was some breathing air. I felt around in the pitch black and knew I was in the engine room. I told myself, "You've got to get out of the water or you're going to have hypothermia and go into shock." I felt around some more and realized I could get myself out of the water completely, so I did.

I started to get super cold. A mattress, clothing, blankets and a sleeping bag floated by. I got back into the water and with all my effort lugged the sleeping bag out of the water, wrung it out and put it around me. The funny part was that I had to pee so bad. I don't know why I just didn't pee my pants, but I struggled to get my pants down before I realized how silly that was. I ended up peeing myself and that's what really warmed me. In the meantime, diesel was coming into my nose, my ears. Every time I touched them, they were just slimy. I could taste it in my mouth. It got into my eyes a bit and every time I blinked they started to sting. All the fumes were making me really sleepy.

Just then, I found this wooden object and started banging away on the hull. I kept banging away, shouting, "I'm alive. I'm alive. I'm down here." Bruce and Kenny heard me and answered, "We're going to come and get you. You just hang in there." But when they didn't show up, I fell asleep. They thought I'd died.

Today, Maxine Matilpi and partner John Livingston create Northwest Coast native art and regalia. Photo: Maxine Matilpi collection.

I must have slept twenty minutes at the most. When I woke up again, that's when reality hit. This wasn't a bad dream. I told myself, "You have to get yourself out of here." So I tried. I got to the washroom and made three attempts to get out of the porthole there. Then I went back to where I was before.

When I got back to the engine room, they yelled to me that four divers were coming through and to stay put but try to be kind of visible, to make a ripple to let them know exactly where I was. So I had to leave the engine room and go back into the water by the stairway. A diver found me and told me that the boat was slowly sinking and we would have to get out within a few minutes. He asked me if I'd ever scuba dived. When I said,

"No," he explained how we would buddy breathe. I was wrapped around him as we were going up. I didn't realize how deep we were. I think we were about twenty, twenty-one feet under the surface.

The boat was upside-down and the hull was still visible. The diver who saved my life gave me a picture of it later. They got me on *Tamanawas*, a boat standing by, and there was a doctor aboard. I was so cold and all I wanted to do was sleep, but he wouldn't let me. Somebody wanted to give me rum or rye and another suggested a hot shower, but the doctor told them either one would give me cardiac arrest. They took off the wet clothing that I had on, put on these huge pants and a shirt, and got ready to hoist me up to a helicopter.

I insisted that the diver who found me come up with me. I hung on for dear life. I remember him telling me not to look down, but I did.

I could hardly believe it, seeing the situation from the air. I had been in the capsized boat for two and a half hours. I had travelled almost the whole of Seymour Narrows going round and round and round. Forest and Betty and the baby were dead.

After the inquest into the accident, I talked to the diver who rescued me. He came up to me and said, "You probably don't recognize me. I'm the one that got you out of the boat." I just gave him a big hug. I got choked up because I wanted to say so many things at once. I thanked him for my life, and he said, "Well, that's my job." I heard that five or six years later he died.

Shortly after the accident, I got pregnant with my son Aubrey and went right back out fishing. But it was with great difficulty, and it was with great pain and fear. Surviving the capsizing has made me appreciate life. I don't take anything for granted. I was lucky to have the two children I've always wanted and that's what made life go on for me. 🐋

Corporal J.C. Lemay was one of three divers awarded a medal for bravery for this rescue effort. Sadly, Corporal Lemay was awarded his posthumously, as he went missing during a training dive and is presumed dead.

Excerpted from *Saltwater Women at Work* by Vickie Jensen, 1995. Reproduced with the permission of Douglas & McIntyre.

SEINING
FOR
HERRING

Previous page: The *Sandy Ann* carves out a set during a 1995 herring opening in the Strait of Georgia. Photo: Michael Skog.

FINDING THE RHYTHM
Origins of the Herring Fishery

Alan Haig-Brown

Today it is the most frenetic fishery on the coast, with prices that make the wildest tales of big sockeye jackpots look mild. But herring has not held such a prestigious place in the BC fishery since the First Nations people of the coast both fished and gathered the roe. When Europeans began fishing on this coast, the native conservation ethic was ignored in the excitement of the development of the river-based sockeye industry. The seine boat was introduced to the salmon fishery just before World War One. Fishermen and businessmen like the Wallace brothers and even big corporations like Delmonte Foods were quick to apply the new technology of gasoline-powered boats and purse seines to the pilchard fishery. Many thousands of tons of these fish were caught for the reduction plants on the west coast of Vancouver Island and "reduced" into oil and meal.

At the same time, herring continued to school in late winter and spawn in early spring as they had for millennia. But Japanese-Canadian fishermen, ever willing to make the best of the less-than-fair position that they had on the coast, began the development of a salt herring fishery in the Gulf Islands. In the earliest years, probably around the First World War,

The *Western Mariner* (left) brails onto the *Sleep Robber* (far left) from a set of about 600 tons made by the *Western Warrior* (right) in Laredo Sound in February, 1961. Photo: Alan Haig-Brown.

FINDING THE RHYTHM

they fished with rowed twin seiners. These appear to have been
based on the classic Columbia
River style of boat with several
pairs of oars and half of the net
piled on the stern of each of two
boats. When a school of herring
was detected, boats would encircle
it in a pincer movement. The net
would then be hand-pursed and

In this 1920s photo, fishermen brail herring into open scows from a seine in the Gulf Islands. Photo: UBC Special Collections.

the herring "brailed" or scooped into scows using "brailers"—large scoop
nets which could be opened at the bottom to release the fish.

By the 1920s the Japanese boatyards in Steveston, such as Kishi,
Nakade, and Atagi, were turning out good, powered seine boats for this fishery. Others such as Ashima at Seymour Creek, and Sawata and Arimoto in
Nanaimo, added to the growing herring fleet. Virtually all of the Gulf Island
herring seiners of the late 1920s were owned by Japanese-Canadians, except
for one fleet owned by a very important Chinese-Canadian fish broker
named Yip who also operated a herring saltery in the Gulf Islands.

The crews of the boats were required by law to be half Japanese and
half European. Louie Percich, who later went on to become a highline herring skipper, remembered working in the Gulf fishery about 1928. They
slept in the fish hold, which was not used for fish as these were brailed
directly into scows. To brail, the men worked with two-handled dip nets as
they sat on the gunwale of the scow with one foot in the scow and the
other in the net. This was back-breaking work that was interrupted one day
by a fisheries officer who ordered the men to line up with whites on one
side and Japanese on the other so that he could count that they were obey-
ing the regulation that required half the crew to be European. As soon as
they had lined up, a problem arose for the puzzled fisheries officer when he
discovered that the crew included two native Indians. All of this of course
provided much cause for laughter by the men who, like all fishermen,
enjoyed seeing the government entangled in their own red tape.

Charlie Clarke, for many years head skipper at Nelson Brothers,
remembered that in the early 1930s it was said that only the Japanese knew
how to locate and catch the elusive schools of herring. He says that they
watched for bubbles on calm days and listened for flipping herring at night.
Then the feeler wire was introduced from Norway. With this approach, a
small skiff was rowed through the water while skilled hands held a weight-
ed wire. As the herring brushed against it, a slight tremor could be felt in
the wire. The thicker the school of herring, the stronger the vibrations in
the wire. As the pilchard stocks began their slow decline from overfishing,
the European-Canadian fishermen turned increasingly to the herring to
feed the reduction plants, which in turn served a growing demand for fish
oil and meal.

With the internment of all residents of Japanese descent in early
1942, the herring salteries fell into disrepair, while the demand for herring

oil went up to meet wartime needs. The fleet fished the whole coast and more reduction plants were built. Growth of the fishery continued into the postwar years. While the Japanese-Canadian fishery had used smaller, twin seiners with very large nets, the new fishery followed the pattern set in the pilchard fishery—a single boat setting on a large school of herring and then brailing across its own deck into a packer or scow tied to the side opposite the net. To keep the seiner from pulling itself into its own net as it pursed, a rowing skiff, called a tow-off boat, was used to keep the seiner away from its net. These were replaced by motorized "power" skiffs about 1960.

Until the introduction of the power block in the 1950s, the body of the net was hauled aboard bit by bit by wrapping a strap around the web and winching a portion, or fleet, of the web with a line running from the deck winch through a pulley at the end of the boom. It was also in the 1950s that advances in electronic sounding technology, particularly the paper recording echo sounder, made it possible for the skipper to locate schools of herring from the pilot house. This was a big improvement over rowing around on cold winter seas with a feeler wire.

In the 1960–61 herring season, the price of herring dropped from $13 a ton to $8.80 a ton, making the use of packers uneconomical. The average seiner could only pack about 100 tons itself, so if a boat got a 500-ton set, which was not uncommon in those times, it was necessary to give most of the catch to other boats. A second boat might tie on the corkline of the boat that had made the set and begin brailing at the same time. If the seas were calm two more boats could tie to the off side of the host boat so that four boats would be working the set at once.

When tightly packed in a "dried-up" net, herring will die fairly soon, and the weight of the dead fish can destroy the net or capsize the vessel, so it is important to get them aboard as quickly as possible. The two-ton brailer with its twenty-foot handle would be swung out into the net. As it hit the water, a "skimmer" line tied to the brailer would be pulled horizontally through the fish. On a signal from the deck boss, who would be holding the brailer handle, another line would be activated to lift the full brailer. As it rose, a third line would swing it across the deck, and a fourth line—the "trip" line—would be yanked at just the right moment to spill the fish into the open hatch of either the host boat or the boat tied alongside. Each line was operated by a different member of the crew. The host boat would heel from side to side as the brailer swung back and forth and the work would fall in time with the rocking motion of the boat as the crew found its rhythm. It was one of the nicest imaginable examples of teamwork on a fishing vessel and made possible the rapid brailing that came to characterize the fishery.

Sadly, the use of high-powered streetlights mounted on the boats to attract the fish—a practice opposed by most fishermen—led to such severe depletion of the stocks that the herring fishery was closed for several years, and only reopened when the roe herring fishery began in the 1970s.

THAT'S FISHING
Seining for Roe Herring

story and photos by Brian Gauvin

It's 1315 hours on March 11, 1988 in Cypress Bay. The roe herring fishery for this area of Vancouver Island's west coast is about to open. Radio traffic in the wheelhouse of the *Northern Dawn* is constant and excited, adding to the tension that has been building since leaving Steveston a week and a half ago.

"They want to take a quick look to make sure, 'cause they're real good fish, real beautiful fish. Then they'll open it an hour or so later," crackles a voice over the air.

"Hey! It's real good roe, real good roe. We've looked at 'em already."

"Do we get more money for 'em?"

The calm water inside Cypress Bay belies the expectant atmosphere. At 1335 hours the bay, located a few miles north of Tofino, holds 35 licensed boats and a number of packers. Their bows point into the bay, they idle and jog about, waiting for the announcement to start sounding.

These are the rules: absolutely no boats, with the exception of the fisheries vessels, are allowed into the fishing area prior to the opening. There is a short period of time the licensed seiners are permitted to scout for herring schools before the area is declared open to fishing—meaning the nets can be set.

The air above, excited by the buzzing, throbbing, and whining of at least four fixed-wing aircraft, a helicopter, and even a low pass by a jet fighter, provides an agitated harmony to the staccato of the radio.

"*Arrow Post* to the fleet. You can start sounding and looking around now." This means permission is given for the large boats to stake out their claims.

"Watch the smoke clouds come now," remarks the *Northern Dawn*'s skiffman, Greg Olafson, as dozens of puffs of diesel exhaust appear simultaneously around the mouth of the bay. The boats head in.

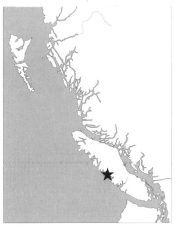

Cypress Bay

The *Northern Dawn* drums up her net.

Brian Gauvin

George Olafson keeps an eye on the sounder. "I'll fish as long as I can hold the arthritis off."

During the long days of waiting, one gets to know a little about the *Northern Dawn* and her crew. The 76-foot seiner, completed by the Matsumoto Shipyard in North Vancouver in 1961, marks the transition from wood to metals by the shipbuilding family. Features such as deck planks that lap over the sheer planks, exceptionally heavy timbers and lock strakes in the decks identify her as a classic wooden Matsumoto.

Gordon Nicholes, the engineer, sums up the feeling of confidence the crew has in the vessel. Pointing out the condition of the caulking and the nails, he comments, "I have to admit, for wooden boats, Matsumoto is the best." He and George Olafson, the skipper, once owned the *Northern Dawn* but sold her to Sid Smith the previous year. The pair attempted semi-retirement, but their experience and familiarity with the boat made it a natural choice for Sid to retain them for herring this season. "I'll fish as long as I can hold the arthritis off," says George.

Still throbbing in line are the original pair of 671 General Motors "Jimmys" which power the main winch, drum hydraulics and move the huge vessel along at seven $7^1/2$ to 8 knots. Sid has added a colour sounder and a 48-mile radar to the wheelhouse electronics. Gordon recalls a trip in the Gulf of Alaska when they were on the fish and the echo sounder quit. "The first thing you will notice is there are two of everything in here," he says, pointing to the dry paper sounder and the 24-mile radar.

Tick...tick...tick...tick. The metronome of the primitive paper sounder marks the seconds as George observes the abstract patterns produced by the herring schools on the colour monitor of the electronic sounder. It is 1430 hours, and with only a few minutes left for sounding George is not the picture of relaxation that semi-retirement should portray.

He first apprenticed at sea at the age of five with his father, Oli Olafson, aboard Oli's gillnetter, the *Laila*. He worked at Oceanic, a cannery in Osland, an Icelandic community at the mouth of the Skeena River, where he grew up in the industry. "I was Canadian Fish's general handy kid from when I was twelve years old 'til I was sixteen, when I started running packers for them."

Greg Olafson, George's nephew, alternately checks the sounder and the other boats in the vicinity. Pointing to a group of five or six boats, he remarks, "They're not jogging around. They're all pointing in the same direction," indicating signs of fish.

Gordon Nicholes sticks his head in the wheel-house, checks the sounder and heads back on deck. For the past eleven days, Gordon, who has spent his life around machinery, has been greasing nipples, checking hydraulic lines, blocks and equipment, making repairs and alterations wherever necessary. Today, it all has to work. "If you are interrupted for any length of time, the fish get panicky. Then you've got problems," he says emphatically.

Tick...tick...tick...tick. The atmosphere is tense in the wheelhouse. A school of about 40 tons appears as a red blob on the sounder. Expecting the announcement to open at any moment, George, looking pale and dry in the mouth, steers the boat in a circle. "Well, there's a little bit to get started on," he says, turning to Greg and giving the order to stand by.

Greg Olafson, thirty-three years old, tall and heavy-set with a red Icelandic beard and a calm, jovial manner, heads for the skiff at the stern. He started fishing with his father, Carl Olafson, on the *Misty Moon* and with George and Gordon on the *Northern Dawn* when he was fifteen.

The tick of the sounder clocks the anticipated opening. "Purse seine fleet in Area 24. This is the patrol vessel *Arrow Post*. Cypress Bay is now open until further notice."

Radio traffic is frantic as George rings the bell, blurts an "okay" and hits the horn. Greg starts the power skiff, kicks it into reverse and unfurls the net off the stern while the larger vessel traces a small circle in the bay. The skiff points in the opposite direction, it now tows the end of the net as a tugboat would tow a log boom.

In the skiff with Greg is twenty-four-year-old Gary Chesal from Nanaimo. He is the second skiffman and deckhand on his first herring trip, but being the youngest crew member aboard may be his toughest job. George completes his circular set and the lines are transferred from the skiff to the seiner. The crew begin pursing up the bottom of the net.

Gary jumps aboard the larger vessel and immediately begins hurling rocks, with sea lion bombs taped to them, over the port side while Greg churns the water with the skiff, in a series of noisy donuts. This racket is meant to discourage the herring from choosing the only avenue of escape, which lies directly under the boat.

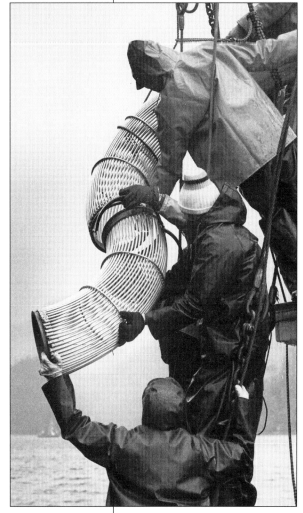

The crew ready the de-watering part of the herring pump. The pump transfers the herring from the net into the holds.

Brian Gauvin

George Olafson helps ready the net for the first set.

Hans Zimmerman, at fifty-five, has twenty years of experience aboard the *Northern Dawn*. Quick to laugh and tell a story, he describes his role as "winchman, deckhand and all round repairman." He is operating the main winch as it hauls in the purse line while Gordon works the drum. For the past eleven days, going over the lines, nets and mechanics with Gordon, he has been fussing with the details that make a difference when things start to happen.

"You've got to be fast in this game," he says.

Tony Smith, the cook-deckhand, was born in Prince Rupert and now resides in Nanaimo. He is talkative, quick thinking, always moving, and throughout the set he is everywhere, relaying information from the drum to the bridge, passing lines, resetting purse rings with Gary—anything to lend a hand. His intensity conveys enthusiasm for a good set.

The herring are boiling at the surface now as the crew begin hauling the net over the side. With Gordon operating the winch, they begin "drying up the net"; using brute force and winches they pull as much of the net aboard as possible to concentrate the fish in one spot beside the boat. Once this is done they can lower a herring pump that sucks up the small fish and pumps them into the holds.

Greg, still in the power skiff, has been holding the larger boat off the bag of herring with a tow line attached to the opposite side of the boat from the net. This becomes more important with larger sets, which can get fouled up with the bottom, or sides of the boat. By gently towing the boat

THAT'S FISHING

away, it allows the net full of fish to balloon out into a nice manageable shape that makes pumping easier.

The first few layers of shimmering silver in the "money bag" are now visible above the surface of the water. None of the crew are impressed with the 35 or so tons and the important thing now is to get them pumped aboard and get some more fish before the fishery closes.

Tick...tick...tick...tick. Darkness is descending on the bay, making the closure of the fishery imminent. An empty second set and watching a colourless sounder have eaten away two precious hours. George steers the *Northern Dawn* toward five or six boats congregating in the shadow of a point near the southern shore of the bay. An eerie glow, cast by a blood red sunset, wraps the scene in an atmosphere of urgency. The boats are quite close now. *Trina Leslie, Western Investor, Snow Queen* and *New Queen* have made sets, leaving little room to manoeuvre. As the *Northern Dawn* noses

Greg Olafson (standing) and Gary Chesal pull the net off the drum to begin the set.

between the boats and skirts their sets, a school of approximately one hundred tons begins to edge its image across the sounder.

"Just five more minutes," says George with disappointment. The area was about to close and there would be no more time for another set. They would be forced to dump whatever it was they managed to corral.

"You've got to be fast, really fast," says Hans to no one in particular.

Tony sums up with "That's fishing."

The *Northern Dawn* slips through the dark night like a small island moving at 8 knots. The journey home along the Vancouver Island coast on a lazy rolling sea is undisturbed. The crew, between watches, sleeps, play cards and talk of the previous days' fishing.

"I shouldn't complain," says George philosophically. "It was a good opening. The fish were spread so that everybody got a crack at them. It's pretty unusual to get a four-hour opening. It's usually anywhere from fifteen minutes to an hour."

That's fishing.

Photos excerpted from *Gone Fishing* by Brian Gauvin, 1995. Reproduced with the permission of Sono Nis Press. Published in the *Westcoast Fisherman,* 1988.

PITLAMPING FOR HERRING

Don Pepper

During the 1960s, I fished in the herring reduction fishery, where the herring were made into fish meal. One of the techniques for catching herring was the use of lights. (We called it pitlamping, after the method for illegally killing deer.) By the late 1960s, the technique was so refined that it threatened to wipe out the herring stocks. Pitlamping was finally banned as being harmful to young salmon smolts, as they also were attracted to the lights.

Prior to this, biologists with the federal Department of Fisheries and Oceans believed that BC's herring stocks could never be wiped out, because it would be uneconomic for fishermen to hunt down the last schools. Therefore, they thought the herring would always rebound with higher growth rates—what they called "recruitment to the fishery." Anyhow, the theory was wrong. Pitlamping was a good technique for catching the last schools of herring, and the technique proved that herring could, indeed, be

Illustration by Graham Wragg

PITLAMPING FOR HERRING

1. The boat is anchored up with the dead skiff and power skiff at the stern. The lights are on, attracting fish.

DEAD SKIFF   POWER SKIFF

2. The boat lights are off and the dead skiff's lights are on. Anchor is up but chain and anchor are still in the water.

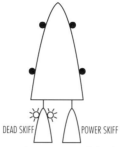

DEAD SKIFF   POWER SKIFF

3. Boat has moved away to set the net while the dead skiff holds the fish.

4. Boat has set the seine net and the dead skiff's lights are out. Power skiff tows the dead skiff out of the net.

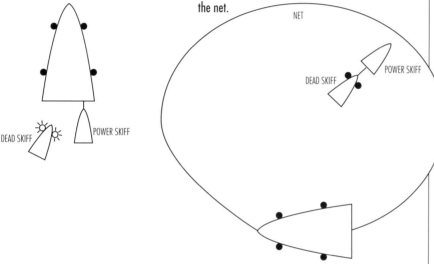

wiped out. As a result, herring fishing was banned until the advent of the present-day roe herring fishery. Nonetheless, pitlamping was interesting in its own right.

To pitlamp herring, you needed a purse seine vessel equipped with strong lights. We used mercury vapour lights—the same as streetlights—and small portable generators. The boat had three or four lights along each side, and a special "dead skiff" was rigged with two to four lights and a gas-powered generator. Thus, the vessel could fish in the normal manner during daylight hours, but could also attract the fish at night. In all, a very efficient fishing method.

In 1966, I fished on the old *Vanisle*, a BC Packers double-decker that packed a nice, even 100 tons. We received $1.10 per ton, per man for herring, so the calculation for a load was easy; my share for a load of herring was $110.

Compared to today, the electronics were abysmal. We had a radio that I always suspected was run by steam-pressure, a Furuno paper-recording sounder, and a compass. There was no radar, no satellite navigation and no sonar, but we still caught lots of fish.

Finding herring in those days was a chore, especially in the summer herring fishery in May and June. The boat never stopped looking. We looked into every nook and cranny on the coast because that was where the herring were. It was the exact opposite of the roe fishery of today, which takes place on spawning concentrations. We were not allowed to do that. We chased scattered, feeding fish. The entire coast was our bailiwick.

During the day, we cruised the 55-fathom line and watched the sounder. A black mark on the paper meant herring. We would immediately turn to starboard and circle back on the school to scout out its size and movement. Easier said than done. To assist us we threw bits of coloured paper or other markers overboard to fix the position of the herring. You had to know where the school was, but when setting your net, you couldn't "touch" them. This meant you wanted black marks on the paper to find them but no marks when setting around the herring. The skill was in figuring out where the herring were and then setting on them so as not to lose them. Sometimes we got them, other times we didn't.

On dark nights with no moon, we pitlamped. We selected a spot by looking for "skim"—light surface concentrations of herring—that indicated abundance, at dusk. Then we anchored at a suitable spot and turned on the lights to attract fish. With the main engine shut down and only the small generators going, it was time for a nap. Any sudden noise was a definite taboo; no slamming doors and no dropping wrenches in the engine room. We waited for several hours.

The power skiff had a sounder and it was sent out about 0300 hours to see if the concentration of fish was enough to set on. We felt 25 tons was worth it, but we really wanted 100 tons. If there was enough to set, we entered the next stage.

The objective was to "hold" the fish to the lights, and then to set around them. First, as noise spooked the fish, the main engine had to be started with as little fanfare as possible. Once the engine was running, the anchor was lifted, but only until the chain was out of the water. The sound of the chain in the chock was bad for holding the fish, so only the cable was spooled; the chain and anchor were left in the water.

The next step was to "transfer" the fish from the boat to the dead skiff. The dead skiff's lights were left on and the main lights on the boat were turned off. This moved the fish from us to it.

After a small pause, the clutch was engaged and, at dead slow, the boat eased away from the dead skiff, which was holding the herring. At a small distance from the fish, the net was set around the dead skiff. If all went well, the dead skiff held the fish in the centre of the net and then the power skiff went to get it and tow it out of the net. We pursed up the net with steel cable purselines and power-blocked the net aboard. If lucky, or skillful, we got 100 tons and headed for Namu, where there was a reduction plant. One load a week was okay fishing; two loads was heaven. If unlucky, we got nothing and went back to the 55-fathom line and tried our luck there.

GROUNDLINE
FISHERIES

Previous page: *Summer Wind* skipper Corey Erikson lands a halibut in Queen Charlotte Sound. Photo: Peter A. Robson.

BLACKCOD DERBY DAYS

Alan Haig-Brown

The catch from the first of the two twenty-five-day openings of the 1987 blackcod (sablefish) season has been delivered. The skippers and crews of most of the 48-boat limited-entry fleet are happy with the results. The fleet is small but diverse. In this fishery, the largest portion is taken by trap boats, followed by longline boats and a small incidental catch by trawlers. In 1986 this was 3,532 tons by trap, 728 tons by longline and 330 tons by trawl.

The trap boats range in size from about 60 feet to well over 140 feet. Doug Beguin fishes the 60-foot *Terry & Gail* in both the prawns and black-cod fisheries. With a relatively small crew he works primarily in the main-

The *Ocean Pearl* is one of only a handful of BC fish boats over 100 feet.

BLACKCOD DERBY DAYS

land inlets and landed a total of about 43,000 pounds of blackcod during the opening.

At the other end of the size scale is Blair Pearl's *Ocean Pearl*, with a registered length of 110 feet. Her twelve-man crew works around the clock on sixteen-hour shifts. This allows three men to be in the bunk at any given time. The boat fishes 850 traps on fourteen strings of six skates each (a skate is a length, or unit, of gear several hundred fathoms long. A string is a series of skates). Each string is 2 miles long for a total gear length of 28 miles. Blair has

Frozen J-cut blackcod. Photo: Peter A. Robson.

used this much gear for the past five years because, "it fits on the boat nice and a little longer soak is better." He says that the minimum soak time—the time gear is left in the water—is thirty hours: any less and the bottom is not being fished thoroughly. In this opening he landed about 370,000 pounds of blackcod, which is frozen on board.

The *Ocean Pearl* catches a fair number of halibut which by regulation the crew are required to return to the sea. But the largest by-catch is undersize (less than 20 inches) blackcod. They too must be returned to the sea. Blair says that, like the halibut, they live just fine if you return them with care. A larger mesh trap would allow some small fish to escape, but would result in a lot of gilled fish that would die before being returned. Blair is concerned about the damage done to incidentally caught fish by the "crucifiers," or automatic hook extractors being used on some longliners. These have been outlawed for halibut but they are still legal for blackcod.

Blair Pearl says that he certainly didn't invent blackcod fishing, but it is commonly said that he was the first to really make a go of it. "I started with pots [traps] in 1977. Before that I longlined for them for years, but never really made any money at it. The first year wasn't any big deal, you could see there was potential, but the market wasn't really there. You could flood the local smoker-market with a couple of trips, a couple of hundred thousand pounds. So we flew to Japan; there's not a big market, but it's a strong market. It's mostly sold in northern Japan in the winter."

The Japanese market seemed to be a steadier market with good prices, but first the Canadian fishermen would have to learn to produce the fish the way that the Japanese wanted. The fish had to be J-cut [a special way of cutting around the head] and the crew had to be retrained. "They didn't really take a lot of interest in us at first. You have to get to know the people and they have to know what you are capable of doing before they take a lot of interest. But we got started and they bought some fish off us. They weren't totally happy with them the first trip, or the second. But it progressively got better and we made several trips over there, trying to learn what they wanted exactly. Finally we got it so that it was pretty acceptable.

In the last few years it became more acceptable than Japanese product and we don't have any trouble selling it."

Blair now has his own marketing firm in Vancouver and the fish are sold to "whoever wants them the most."

While this opening has been "short and sweet" for Blair, it is not the way he would like to see the fishery managed. "The way it is set up this year, with the split season [two openings at different times of year, where all the year's catch is taken], is not too desirable for a lot of the fleet. It makes it so damn expensive. You have to change all your gear back and forth."

It takes a full month to change gear on the *Ocean Pearl* and with two openings this has to be done twice a year. The blackcod fishery has since changed over to an individual quota system, where each vessel is licensed to catch a certain tonnage during an extended season, but in 1987 Blair and others are weighing the plusses and minuses of moving to quotas. "I don't really like it, but it's better than anything else that's available right now. It would allow you to forecast what you are going to do for the next year a heck of a lot better. Everybody would realize more money on the fish through better marketing. It would cut your overhead down in terms of gear. We pack so damn much gear now and we haul gear in such lousy weather that we are tearing it apart. It's just too competitive. If you're not on a twenty-five-day opening you'd wait until the blow went down at least, but now you don't stop for anything."

Don Quast, on the 84-foot *Milbanke Sound*, fishes 470 traps. He agrees with Blair on the advantages of quotas. "It would be good if it worked out. Then we could take advantage of the prices. You wouldn't have the whole fleet on the ground at once and you could fish in February when the fish are still in shallow water."

He went on to explain that on this last opening they had fished as deep as 450 fathoms and in the summer they have been down 900 fathoms, but in summer you could get the fish at 250 to 300 fathoms.

Blair Pearl thinks that the research on the blackcod fishery is going well and that the biologists are getting a better understanding of the fishery. His catch per trap is going up a little every year as is that of the fleet, but he's not sure whether this is due to better fishing techniques or an increase in the stock. He used to prospect around for fish during the season. But now "you can't do it any more. After the first hour of the opening, the whole coast is covered with gear. We pick out an area and as long as there are a few fish there we keep plugging away. But you can't fish back over the same ground."

Blair Pearl isn't one to hang around on the same grounds anyway. He's a progressive fisherman who stays open to new ideas and new fisheries.

RUPERT LOOKS BETTER OVER THE BOW

Halibut Derby Days Aboard the *Dovre B*

story and photos by Brian Gauvin

Hecate Strait

"Help!" yells Gary Robinson. Crimson rivulets of blood run down the side of the hatch, disappear under a full pen of fish carcasses, mix with fish slime and seawater and pour red out of the scuppers. Al Newton, up to his knees in slippery halibut, reaches a helping hand over the rail and secures a hold on the big "soaker," a nickname for a very large halibut. Gary Robinson has a gaff embedded in the behemoth's head, his left hand yanking on the gangion—the short length of line attaching the hooks to the ground line. Together they wrestle the halibut over the bulwarks and into the checker.

The crew of the *Dovre B* are dressed in rain gear as protection against the early morning mist, blood, fish slime and the wash of the sea as it streams over the lurching deck. A full checker and more fish coming over the roller sharpen the crew's spirits after two and a half days of monotonous labouring for a meager 2,000 pounds of halibut.

"There's nothing easy about halibut. It's all hard work," says skipper John Newton.

John, at the remote controls, is keeping the boat in line with the groundline, but is close enough to assist Gary in gaffing. There is a fish on

The crew of the *Dovre B*—Ken McDonald, Al Newton, John Sylvestre, Mickey Pilfold, John Newton and Gary Robinson.

almost every hook now as the crew settles into a working rhythm. Gary slams a fish into the pen, then it's back to the rail, poised for the next offering seven seconds later. Al clubs a number of thrashing halibut, hoists them onto the hatch and begins dressing them. Hunched over, Mickey Pilfold coils fathoms and fathoms of groundline as the gurdy hauls the gear aboard. Ken McDonald chops cod and pink salmon into chunks. And John Sylvestre is in the stern, piercing the chunks of bait with the circle hooks

that are attached by gangions every 18 feet onto the groundline.

And so it goes. Day after day rolling into night and back into day.

It is August 22, 1988—in the days before halibut fishermen were assigned individual quotas which they were free to fill at their leisure, anytime during the March to October halibut season. In the pre-quota days, openings were short and intense. Boats had to get on the fish fast and fish them hard to make the season pay.

John set the gear in the middle of Hecate Strait at a spot that last year yielded very good results for the late summer opening. This year, in the same spot, 250 fathoms of gear was intercepted by a layer of dogfish before it reached the halibut, feeding at the bottom. The unwanted bounty taxed the roller-men's energy, sapped the crew's spirits and forced a time-consuming move to another spot.

Working late into the night and starting early the following morning, the crew haul all but two of the eight strings. They begin "farming" another area with an exploratory pattern of gear laid in search of thicker densities of the fish. By the afternoon of the second day, all of the gear has been moved over to the new spot. Lingcod and red snapper have replaced dogfish, and one string excites everyone by producing a number of medium-sized halibut. John decides to set two strings on end and give them a long soak overnight. If they produce, we'll stay and farm this area. If not, then it's back to the first spot to do a test

RUPERT LOOKS BETTER OVER THE BOW

set and see if the dogs have moved off. After two very slow days of fishing, the mood aboard is positive, but far from enthusiastic.

"You can sure tell when fishing is bad. There's no horseplay and joking around," comments John.

The next morning the mood has improved.

"That's the way we like to see 'em! I wonder if we've got them all out of there?" John says, after they pull in the last of a very productive set. As soon as all of the gear is in he heads for the wheelhouse, notes their position from the Loran co-ordinates in his halibut log book, lays on the throttle and heads back over the ground they have just covered. Ken fires the slingshot (a line tied to an anchor at the end of the buoy line) back to Al at the stern. Al ties it to the groundline and yells "Ready in the chute!" Ken throws the flag pole and the bladder over the rail and hurls coils of buoy line after them. Hoisting the anchor onto the rail, he shouts "On the rail!" and drops it over the side when the slack is out of the buoy line. The groundline, complete with gangions that hold the hooks, slowly unfurls till we reach the other anchor, and that too goes over the side with a flag pole attached. John then turns the boat 90 degrees, runs for three minutes, turns another 90 degrees and brings the boat back over the hole.

When the gear has soaked for long enough, it is hauled aboard by Al who gaffs each halibut and flips it aboard in such a way that the hook rips free from the mouth of each fish. This 2,000-pound string of halibut has made Al "high dory"—a reference to the days when individual fishermen fished from small wooden dories that delivered to a mother ship.

The pens are full and the hatch is covered in blood, fish and chunks of bait. John steps up to the roller for his turn on the next string. Halibut after halibut comes aboard, and John, with sweat beading on his face, yells "I ain't fifty any more!" as he hauls aboard another fish. When the 2,500-pound show is over, John displays a big grin and declares "Now I'm high dory."

Gary Robinson has a tough workout at the roller on a big string.

Brian Gauvin

Skipper John Newton gaffs a halibut into the checker while Gary Robinson dresses fish on the hatch.

Even though it's after 1700 hours, John decides to set another two strings. "I've been told time and again by old-timers never to set after five," he says. The chances of starfish getting the bait are increased at night. Yet everyone feels that this spot could save the trip, therefore it warrants a gamble.

Surrounded by the black night, deck lights capture the spray from fish flung into the checker or heaved onto the hatch for dressing. The scene resembles a small factory during a graveyard shift. Gary deftly cuts around the gills, rips them out, flings them overboard, slices down both sides of the blood line, yanks out the intestines, sends them chasing after the gills and scrapes the sweet-meat and the blood from the cavity. In a small pool of blood on the hatch cover, a halibut heart has been beating for over twenty minutes. Gary digs out a tiny halibut ear and sets it down beside the beating heart, pointing out the minuscule growth rings that indicate the age of the fish; he can tell the good years from the bad by the width of the rings, just like on a tree. He has counted thousands of these growth rings during four summers working for the International Pacific Halibut Commission (IPHC) while gaining his biology degree. Grasping the fish by the tail and where the gills used to be, he flips it into the pen behind the hatch where it will stay until put to bed in the hold. The heart is still beating.

John also has an interest in the Commission. As the IPHC representative for the Prince Rupert Vessel Owners' Association, he feels that, of all the bureaucratic agencies involved in the fishing industry, it manages to do particularly good work.

"In my opinion, if we didn't have the Commission there wouldn't be a halibut left in the ocean."

A thin white strip of fog splits the blue water from jagged black mountains to port and starboard. This morning, in the brilliant sunshine and crystal clear air, we can see both the mainland and the Queen Charlottes. The late set last evening has come up empty, lending credence to the "after five" theory. The crew move through the endless rotation of roller, checker, coiler and back to the claim (where the hooks are baited) to prepare the strings. A cheer goes up as Ken one-hands a halibut into the checker on one of two good strings today. The fishery closes tomorrow at noon.

"This is the last get up," sighs Al on the morning of August 25. All of the gear is in the water and John hopes to get two more strings in quickly in order to get it all aboard by noon.

"We started with dogfish and we're ending with dogfish," concludes Al, watching one after another come over the roller. There is just over 20,000 pounds of halibut, plus mixed fish in the hold when the *Dovre B* begins a twelve-hour run to the Co-op in Prince Rupert, the former halibut capital of the world.

"I never said I'd never go halibut fishing again, but I've said lots of times I *wished* I was never going halibut fishing again," remarks John.

"Rupert looks better over the bow than it does over the stern," says Al.

Photos excerpted from *Gone Fishing* by Brian Gauvin, 1995. Reproduced with the permission of Sono Nis Press. Published in the *Westcoast Fisherman*, 1988.

Queen Charlotte Sound

The Eriksons' *Summer Wind* is a 55-foot wooden Frank Fredette design built by Cyril Thames in 1974 at his yard in Bowser. She is powered by a 671 Jimmy.

SIX DAYS ON THE *SUMMER WIND*
Halibut After Quota

story and photos by Peter A. Robson

Sunday, April 24, 1994

1730 hours. After departing Sointula earlier this afternoon, the handsome 55-foot Fredette troller/longliner *Summer Wind* motors past Port Hardy and up Goletas Channel towards Nahwitti Bar. It will take another twelve hours to arrive at the halibut fishing grounds in the open expanse of Queen Charlotte Sound. Aboard are skipper Corey Erikson and crew members Dale Erikson (Corey's brother) and Todd Vick. The three crew range in age from twenty-three to twenty-five years, but their youth belies an ocean of experience, both learned and inherited.

All three come from long-established fishing backgrounds. Corey and Dale's great grandfather Gustaf Erikson emigrated from Sweden in 1917, four generations ago. Corey, Dale and brother Wesley started fishing with their parents as children and began earning wages at seven years. They were fishing on their own at sixteen. At various times the family has trolled, longlined, trapped, dragged, seined and packed. Corey is now running his father Ken's *Summer Wind*, Dale runs the *Majestic Belle*, a 41-foot Deltaga troller/longliner and Wesley runs the *Rosalie I*, a 41-foot troller.

Todd Vick got his first taste of the sea during grade school when his father, who worked on a salmon packer out of Prince Rupert, would take him aboard for the summers. Since then, he has spent two years at college, and has trolled salmon, longlined halibut and tried his hand at dragging. Todd first met up with the Eriksons in Mexico. He got along well with the family and ended up fishing with Dale in 1993. Says Todd, "I like to be outdoors. That's why I'm in the business." When not fishing, he is an avid wilderness camper, hunter, and sport fisherman. Many of Todd's family are also in the business. His uncle John is part owner of the well-known marine

supplier Vick Enterprises, his uncle Ed Skog has a long history with J.S. McMillan, a cousin seines, and two of his mom's brothers were halibut fishermen.

The *Summer Wind* fishes two halibut quotas—the maximum any vessel can fish of this species under the individual quota system—totalling 74,000 pounds of fish. The vessel and her crew have made several trips since the March 1 opening and 34,500 pounds of quota are still to be fished. Corey hopes to put in 20,000 to 25,000 pounds this trip and fish the remaining quota after he winds up the salmon trolling season aboard the same boat.

Monday, April 25

0600 hours. Time to set the gear. According to the *Summer Wind*'s plotter, we are on the grounds. No land is in sight and Corey says we won't see any until we head back to port. Overnight, the westerly winds have increased to over 30 knots. The sea is sloppy with lots of cresting whitecaps. The stabilizers are down, however, and doing a good job of dampening our roll.

The flag and buoy for the first string are tossed out. Corey calls for 250 fathoms of buoy line. Todd uses the spread of his arms to count the fathoms as the line spools off the drum. When he shouts "250," Corey slows the boat. An anchor is tied to the groundline and tossed overboard.

The *Summer Wind*'s crew members range in age from 23 to 25, but their youth belies an ocean of experience.

With one end of the string anchored and marked by a buoy, Dale and Todd climb into the checkers at the stern and begin the tedious process of baiting the hooks with octopus chunks and clipping the two-foot-long leaders, called gangions, to the groundline every ten feet. They work together like a precision drill team, never missing a beat as the gear streams out behind the boat.

Watching his plotter carefully, Corey steers along the plots he has programmed into the computer. When he reaches the end of his marks, he pulls back on the throttle. Dale and Todd tie in another anchor and toss it overboard. Corey pushes the throttle forward while over 200 fathoms of buoyline are spooled off the drum. A flag and buoy are attached, the groundline is cut, and the works is tossed overboard.

Over the next couple of hours, two more strings are set. Each is parallel to the others and about half a mile apart. Corey sets the autopilot and the modest-sized wooden vessel idles into the swell. After a quick breakfast, the crew retires for a few hours of sleep.

The wind has dropped to about 15 to 20 knots, but the sea remains sloppy. The sky is mostly overcast, with the occasional sunny break. Albatross float lazily alongside. Three other longliners are on the horizon. No land anywhere.

1430 hours. The gear has been soaking for about six hours. Corey pushes the throttle forward and steers towards the first string. A blustery 30-knot wind is still blowing sloppy seas in from the west. The sky is now clear and blue.

Taking a break—Dale Erikson, Corey Erikson and Todd Vick.

1500 hours. The flag and buoy for the first string are wrestled aboard. The buoyline is tied to the drum, and Corey takes up the rollerman station amidships, where he will work the longline gear, the drum and autopilot controls. Dale and Todd are suited up and ready at their work stations on the hold hatch cover with knives and scrapers. Penboards have been fitted across the deck to hold the catch.

Within moments, Corey is swinging flopping halibut aboard. Each gangion is quickly unclipped from the groundline and the empty hooks are racked. The few dogfish, skates and blackcod that are hooked are returned to the sea. A small percentage of rockfish, mainly idiots (shortspine rockfish) and convicts (Redbanded rockfish), are caught and retained; a 20-percent (by weight) bycatch is permitted in the fishery.

Meanwhile, Todd heaves the halibut from the deck onto the hatch covers. He measures the smaller fish against marks on his cleaning mat and releases any fish under the 32-inch minimum size requirement. With a well-practised hand, he gills and guts each legal-sized one, then slides it over to Dale. Dale reaches into the stomach cavity and removes the sex glands, then uses a combination hose and scraper for the final cleaning. Cleaned fish are dropped into an enclosed area on deck.

For the next six hours, the work continues purposefully. The fishing is good and the mood is upbeat. The three crewmen keep up a running dialogue and the conversation covers the price of halibut, memorable vacations and past and present love interests in the three bachelors' lives.

2045 hours. The last buoy is aboard. The three strings brought in 231 halibut with a total weight of about 4,500 pounds. The average weight is estimated at around 20 pounds.

The wind is now up to about 35 knots. Corey sets the autopilot and we jog into the lumpy seas for the balance of the night.

Tuesday, April 26

Corey had planned to set the gear out again at 0400 hours, but the wind is blowing 40 knots from the west, making it too rough to set safely. It is interesting to note that one of the reasons for the switch from derby-style, or shotgun, openings to a quota system was to protect fishermen from having to fish in hazardous weather conditions. Nevertheless, we are still twelve hours from a safe harbour, and it is not practical to run from the weather. "Anyways," shrugs Corey, "it always blows out here."

Corey says the worst weather he has ever faced was at the opening of the 1993 season. In company with Dale on the *Majestic Belle*, the *Summer Wind* had left Bella Bella on a Friday (perhaps their first mistake, for many consider that departure day as an ill omen). The boats soon found themselves in an 80-knot southeasterly. They rode out the weather for twenty-four hours, during which time, Corey says, quartering waves were washing right over the cabin. The worst moment, Corey recalls, was when the ten-by-ten-foot main hold cover washed away. Both boats survived, and although Corey admits he was afraid of dying at the time, he shrugs, "It's just part of the job."

0600 hours. The winds have dropped to 30 knots and Corey decides to set. With only slight modifications, the gear will be set in the same location as the previous day. Again, Dale and Todd stand by the

Tossing the anchor for the groundline.

Dale and Todd bait the hooks manually and clip them onto the groundline.

The groundline has soaked for six hours or so. Crew stand by to retrieve a buoy.

checker and bait the hooks. In the wheelhouse, Corey follows the marks on the plotter.

Many longline vessels have gone to automatic baiters, but the *Summer Wind* has stayed with the traditional but slower method of hand baiting. Corey explains that under the former derby-style fishery it was important to get your gear out as fast as possible and lay as much of it out as you possibly could. Now that the fishery is under a quota system, the pressure is off and he sees no need to automate. He also feels that fresh bait is vital in maximizing catches. This belief is grounded in the theory that cutting and using bait as soon as it thaws is better than having it sit in tubs for long periods of time, as would be the case in pre-loaded bait for automatic systems.

After the three strings of gear are baited and set, we idle into the swell again. The weather forecast is for easing winds then a switch to moderate southerlies. Corey cooks up a deluxe eggs Benedict breakfast, then the crew watch a video before climbing into their bunks for a few hours.

1800 hours. Everyone is rested and ready to haul the gear. As predicted, the westerly has eased to about 20 knots during the afternoon. The first string looks pretty good, and plenty of halibut are coming aboard. The gear hangs up several times, but is cleared without mishap. One hundred and six halibut are brought aboard on the first string. The second string brings in 87 fish. As evening turns to night, the deck lights are switched on, and the temperature turns biting cold.

Wednesday, April 27

0300 hours. The last of the halibut caught on the third string have been cleaned and iced. The wind has eased to about 10 knots and the seas are calming. The blood-soaked crew is wet and weary. With the third string, the boys hit pay dirt. One hundred and twenty halibut, averaging 25 pounds, were landed. The 3,000 string is short of Corey's best-ever string of 5,000 pounds, but certainly one to be pleased about. Corey estimates that there are now approximately 11,500 pounds of halibut on board. It is not long before the crew is climbing into bed for a few hours' sleep.

0745 hours. Back at her again. Three strings are set over the next three hours. The albatross seem to have figured us out. They sit patiently behind the boat and try to pick off the bait as it streams out. The day is dawning beautifully. There is a light haze, but sun shines through. The wind is steady at 10 knots. The crew proclaim this the calmest day so far this season.

1800 hours. The crew is up and around. The weather forecast is for moderate to strong southeast winds. Surprisingly, Triangle Island is visible as a bump on the horizon, even though it lies some 25 to 30 miles away.

SIX DAYS ON THE *SUMMER WIND*

Clouds are building on the horizon. Looks like wind is on its way. Four other longliners are in view.

1900 hours. The crew start hauling the gear. Corey is suffering from severe back pains. Dale takes over as rollerman, and Corey scrapes. The first string comes aboard with 93 fish. One of those, however, slaps the cockpit's autopilot controls with its tail and tears the plastic faceplate off. Corey and Dale spend forty-five minutes rewiring a spare. Hauling continues.

Thursday, April 28

0100 hours. The second string is aboard. It yields 80 pieces with a 22-pound average.

0415 hours. Third string is in. We hit pay dirt again. One hundred and twelve nice fish are landed. They average 28 pounds each. The string has brought in about 3,000 pounds. Approximately 19,000 pounds are now aboard. The tedious job of icing down follows. Corey decides to set two strings over the most productive area right away, then head back to town after they are hauled. The weather stays calm, but a storm is definitely brewing.

0600 hours. The last string is set. Dale and Todd head off to bed. Corey takes the wheel and decides to run east for two hours so that he can use the Autotel to check on prices.

1130 hours. We arrive back at our gear. Another longliner is very close to our gear. Its skipper radios us, saying that he had thought we had gone to town. Thinking us gone he had "inadvertently" set his gear very near our own. He hints that if he has set too close, he will come and haul his gear. The crew of the *Summer Wind* grumble. The two sets of gear are very, very close together.

A gale warning for a 55-knot southeasterly comes in over the radio. Currently, though, the winds are only about 20 knots.

Corey hauls 'em aboard (centre) while Todd (right) guts and gills them. Dale then removes the sex glands and does the final scraping.

Things don't always go smoothly. This tangle is the result of another boat setting its gear over the *Summer Wind*'s.

1500 hours. We begin picking up the first string. As we do so, the new autopilot control decides to "pack it in." Corey and Dale go to work again, pulling it apart, and checking all the connections. Half an hour later, it is working again. Gear pulling continues. But, almost immediately, more trouble strikes. The other boat's gear has been set over top of ours and no amount of manoeuvring can free it. Corey gets on the VHF and radios the other boat to come haul their gear. Before he arrives, however, our gear starts coming aboard again. The other boat is advised to stand down. Hauling continues, but the catch is poor.

1630 hours. We are hung up on the other boat's gear again. This time, our groundline breaks under the strain of trying to clear it. We will try hauling from the other end. Corey decides to get the other string first. By now, the crew just wants to get done and head home. The second string yields 104 halibut.

2055 hours. The first broken string is hauled aboard as a tangled mess. It is time to turn around and buck back to port. In our hold are over 20,000 pounds of halibut and about 1,000 pounds of rockfish.

Friday, April 29

1000 hours. The expected gale has missed us, even though winds in excess of 55 knots are reported on the North Island. We arrive without incident at Port Hardy. Within moments, Corey is up at the pay phone calling buyers, trying to get the highest price. With luck, he'll be able to sell the catch wholesale for about $5 per pound. 🐟

TRAWL
FISHERIES

Previous page: Emptying rockfish from the cod end of the F.V. *Caledonian*'s trawl net. Photo: Peter A. Robson.

ON BOARD THE F.V. *CALEDONIAN*

Peter A. Robson

The needle of the wind-speed indicator shows gusts to 60 knots. The *Caledonian* shudders as she tows her trawl across the ocean bottom 150 fathoms below. The sea is white, streaked, and wild. Spume fills the air and mountainous seas break over the bulwarks, flooding the deck with seething torrents of icy November seawater. The wind has steadily increased throughout the afternoon and is now peaking. Darkness is almost complete. Skipper Allan Marsden has just announced that once the net is aboard we'll shelter for the night at nearby Kunghit Island, in the southern Queen Charlottes. Finally, the ship's horn sounds a long mournful blast to signal haul-back time.

Photo: Marsden collection.

ON BOARD THE F.V. *CALEDONIAN*

A full cod end pops to the surface. Photo: Peter A. Robson.

Donning thigh-high boots and orange, yellow and green rain gear, the crew make their way to the stern. Meanwhile, in the wheelhouse the skipper has engaged the two winches to haul in the 400 fathoms of inch-thick steel cables that make up the main warps. Standing aft, the crew are submerged repeatedly, and it's difficult to see where Hecate Strait ends and the *Caledonian*'s rust-stained turquoise decks begin.

A full fifteen minutes pass before the steel trawl doors—used to spread the net underwater—surface and are stowed into place. The crew connect the heavy chain transfer lines from the net drum to the trawl ground warps and the tension is taken up on the drum. Slowly, the net is wound aboard till the cod end pops to the surface behind the boat. When this bag of fish lies at the bottom of the stern ramp, a heavy strap is cinched around the empty part of the web preceding the catch and connected to a cable stretching from a winch on the gantry, the large crossbeam aft of the wheelhouse. The tension is taken up on the strap and with the drum unwinding, the cod end is winched up the stern ramp and along the well that runs the length of the deck.

With the net aboard, the skipper increases the throttle, and we change course for protected waters. Crew member Fred Marsden releases the puckering knot in the cod end and the fish begin pouring out into the well. Additional straps are rigged, then raised in such a way to allow gravity to empty the net of its fish. Several rips are quickly mended, and the net is spooled back on the drum. The hold is opened and the crew, standing thigh-high in fish and swirling seas, begin culling the fish. All told, the haul-back is estimated at 10,000 pounds—a relatively small catch of orange-coloured rockfish. The *Caledonian* has finished her third day of fishing. Her holds now contain about 70,000 pounds.

The 115-foot steel trawler and her five-man crew left Vancouver four days ago, on November 4. This ten-day trip is the last scheduled for the

Hecate Strait

year. Upon her return, the vessel will be drydocked for minor repairs and will undergo her four-year Canadian Steamship Inspection.

Built in 1974 by Benson Brothers of Vancouver, the *Caledonian* is one of approximately ten stern-ramp trawlers working the BC coast. Her six refrigerated saltwater holds can carry close to 400,000 pounds, and keep rockfish for up to nine days. She fishes for J.S. McMillan Fisheries and is partly owned by skipper Allan Marsden. The fact that three of her five crew members have been aboard from four to fifteen years shows that the *Caledonian* is a good boat to work on. Not only does she hold a steady crew, but over the years she has proved herself extremely seaworthy and boasts an impressive catch record. This vessel has spent its past years trawling for groundfish, packing herring in the late winter, and catching hake for foreign factory ships during the salmon season when domestic processing plants are at their busiest.

There are about thirty species of rockfish in BC's coastal waters. Among those aboard are ocean perch, convict fish, rougheye, idiotfish, silvergray, chilipeppers, splitlips, darkblotches, canary, yelloweye, redstripe and harlequin rockfish. A small amount of cod is also being caught. Species include lingcod, greencod, browncod and blackcod. Allan explains that the valuable blackcod fishery had its beginnings with the trawlers. Now, due to the growth of the blackcod industry through longlining and trapping, all trawlers are restricted to a trip quota that is set before each season. Allan doesn't target this species, and their catch is entirely incidental. Later in the trip, soft fish such as turbot, sole, flounder, pollack, brill, witches (rex sole) and skate will be targeted to top off the holds. These are usually pursued near the end of each trip for they do not keep as long in the holds, and they are not as valuable.

Skipper and crew come from across the country, bringing a variety of different skills. Like many west coast trawler crews, the money is good, they work well together and turnovers are infrequent.

Allan began his fishing career seining for herring out of Newfoundland. It was this work that gave him the opportunity to work with fishermen from BC, including skipper Peter Newman. The fishing back east was poor, so Allan decided to move west. His original thought was to head to Great Slave Lake, but Peter talked him into heading to BC instead. Allan made his way to Prince Rupert and found work on George Warner's salmon seiner *Betty G*. At the end of the salmon season, Allan needed to find other work. His friend Leonard Smith introduced Allan to David Clattenburg, part owner of trawler *Pacific Rover*. Despite knowing very little about trawling, Allan was hired. A year later he was skipper of the trawler *Taplow*. A year after that, at the age of twenty, Allan had his master's ticket and was on his way to becoming one of BC's most successful trawler skippers.

Engineer Allan "Chief" Newfeld worked on packers and fished blackcod before joining the *Caledonian*'s crew in 1980. In addition to tending the ship's engines and other systems, he spends much of his time opening and closing valves, and pumping seawater through the seven chillers, one hold at a time, to keep each hold between 29.5 and 30 degrees Fahrenheit.

Gene Power has been aboard since the boat was new in 1974. A former Newfoundlander, Gene came out to this coast for a holiday in the mid-sixties and is still here. Both he and Chief Newfeld were aboard when Allan Marsden became a partner in the *Caledonian*. Says Gene, "I was sold with the boat—I just hope they got a good price for me."

Mate Fred Marsden, like his brother Allan, also headed west from Newfoundland. Fred owned a contracting business in Newfoundland prior to coming out west. Wanting a change, he decided to try fishing. Fred worked on the *Arctic Harvester* for two and a half years before switching over to the *Caledonian*. Two missing fingertips attest to the hazards of working on trawlers.

Former Nova Scotian Wayne Allen is serving as net-man, filling in for regular crewman Sid Dory. Previous to signing on the *Caledonian*, Wayne worked back east as a trap and gillnet fisherman, rigger, net maker and designer.

Day Four

Winds have calmed since last night and we're back on the grounds. Worse weather is predicted tonight, but we hope to get in a day's fishing between blows. Trawler crews soon become used to fishing in bad weather in exposed waters. The distance they fish offshore often makes it impossible to run back into sheltered waters. Although the trawl is occasionally worked in winds of 60 knots, handling the heavy gear when offshore generally becomes too dangerous at about 45 knots. Allan explains that the *Caledonian* has been caught offshore on several occasions when wind speeds have reached 100 knots. Under those conditions his only strategy is to jog into the weather.

Around the galley table, left to right: Allan Marsden, Wayne Allen, Fred Marsden, Gene Power and Allan "Chief" Newfeld. The TV above their heads is well tied down with wire. Photo: Peter A. Robson.

Up in the wheelhouse, Allan plugs a cassette tape into the colour video plotter. Previous trawl tracks are represented by coloured lines winding across the screen. Some have red symbols showing snags, turning or reference points. The Loran land-based navigational signal network enters the boat's position as a waypoint every minute, and the position shows on the screen as a flashing blue diamond. Two colour echo sounders show the shape of the bottom and any feed. A paper sounder is connected to a second Loran, but is used only for mid-water trawling.

Considering the time she spends offshore, Allan has equipped the *Caledonian* with a 48-mile radar as well as a 72-mile radar. A cellular phone, three VHF radios and two SSB radios keep the huge trawler in touch. One of the most comforting features in the wheelhouse is a deluxe helmsman's seat, complete with an adjustable lumbar support. This is important as Allan spends the entire daylight hours, and often much of the evening, on station in the wheelhouse. The crew generally stand the night watches.

Soon after Allan arrived out west his success in skippering trawlers led to ownership. It was Vagn Mark, part owner of the trawler *Pacific Rover*, who gave Allan his first real break, helping finance Allan to purchase shares in the *Taplow*.

A short while later, Allan became a partner with Vagn in the trawler *Scotia Cape*. Part owner Tommy Foote was running the *Scotia Cape* most of the time while Vagn ran the *Taplow*. Allan spent his time as relief skipper until he was hired as skipper on the trawler *Royal City* for Canfisco (Canadian Fishing Company). After a few years, the *Taplow* was sold and the *Scotia Cape* was lost without a trace, so Allan and Vagn decided to dissolve their partnership. Allan then became the major partner in the troller *Pat T II*. Although Allan never fished the boat, the partnership didn't go well and the experience almost bankrupted him.

About the same time, Canfisco sold their bottom fish operation and the *Royal City* was tied up. Allan continued to work as relief skipper for other vessels. In 1984, Allan became skipper on the trawler *Caledonian*, which fishes for J.S. McMillan Fisheries, and was soon offered a one-third share in the boat. The partnership proved a productive one, and three years later, Allan was also part owner of the trawler *Arctic Harvester*. Today, Allan spends most of his time running the very successful *Caledonian*. After the first two years of operation, these two vessels have caught approximately 10 million pounds of fish. Says Allan, "You just never know, it's working now, but it could all be lost in a minute and I'd be back to square one."

During this day, three three-hour sets are made, adding only about 10,000 pounds of fish to the holds. Asked if he spends much time searching for new fishing areas, Allan answers, "It's not really worth experimenting with new grounds as rip-ups and snags, combined with the low quotas, make it generally not profitable to do so. Enough fish are around to make quotas on known grounds."

Allan explains that the strategy for trawling has changed significantly over recent years. Mixes are being targeted, not quantities of individual species. This has made the trawler skipper's job more difficult. Says

Skipper Allan Marsden at the wheel. Photo: Peter A. Robson.

ON BOARD THE F.V. *CALEDONIAN*

Allan, "There's plenty of fish out there. Finding them isn't the problem, it's finding the right mix to meet quotas that's difficult. Otherwise I wouldn't even be fishing here. It's no longer worth using the fuel to get to the more productive areas for certain species because the quotas are so low." The skipper would ideally fill the boat in such a way that the non-quota species provide the bulk of the catch while the quota species are caught to quotas. "For example," explains Allan, "if the trip is allowed a hundred thousand pounds of quota fish, you hope to end up with an additional two to three hundred thousand pounds of incidental fish."

Day Five

Today, we are using the soft-bottom net and towing it at the standard 3 knots. The *Caledonian* carries three trawl nets: a soft-bottom one (Engle trawl), a hard-bottom one (western II or box trawl), and a mid-water net. The mid-water trawl is used mainly for the times when the *Caledonian* fishes for hake in the summer, or on occasion for pollack, greenies and brownies. When this net is to be used, the cod end and intermediate from one of the other nets is laced on. Gene Power explains that

A chain knot holds the cod end closed against the strain of tons of fish. Photo: Peter A. Robson.

the process only takes about fifteen minutes. Trawl nets are by far the most difficult of any to construct. Up to ten different mesh sizes may be used and each net may have as many as two dozen prefabricated mesh panels, many of them tapered. In addition, the nets have chafing gear covering the bottom and sides of the cod end. Each net has a different assortment of weights, floats and bobbins, depending on bottom conditions and owner preference. The cost of trawl gear—including net, doors, and cables—can exceed $40,000. The day's first haul-back nets about 9,000 pounds. Once again the catch is a good mix, according to Allan, of one-third quota fish and the rest non-quota. The second set nets about the same quantity.

During the afternoon we see about twenty humpback whales travelling west, some as close as 100 feet from the boat. Hauling back the third set we find one of the one-inch-thick steel bridle cables has broken. It shows the tremendous strain the gear takes. Allan points out that the cables are trimmed off a couple of times per season because of wear from dragging along the bottom. Within fifteen minutes a spare bridle is rigged. However, the damage is done and the haul-back nets a mere 1,000 pounds of fish. So far, virtually every set of the trip has torn up the net to some extent. These rips are quickly mended before the net is shot out again. The net repair needles are never far from the crew's reach.

We spend the night jogging further south in the Straits. The skipper talks over the VHF to the *Arctic Harvester* which has just arrived on the grounds. Her navigation lights are showing several miles away as she, too, jogs for the night. Allan guesses there are about twenty trawlers working the coast at any given time.

When asked about the dangers inherent in handling spiny rockfish, the crew agree that although they do get poked and scraped on occasion, injuries are easy to avoid if they're smart enough not to actually kick the fish. Rockfish spikes are not poison, per se, although it's possible to get fish poisoning from them. Ratfish, however, which are often in the trawl, do have poisonous spikes on their tails. As the fish are shovelled or pushed into the hold, they are seldom actually handled. Even so, it is amazing to see the crew wading thigh high in the well and not getting speared.

Day Six

The skipper has decided we are getting close to quotas on some species. As a result, we have shifted to an area Allan hopes will yield up the balance of his quota species and top off the holds with a big catch of turbot. Although the price is much lower for this small, flat fish, it has no quota, and with luck he can make up for the low price by volume.

As turbot, unlike most of the rockfish, can be caught in darkness, the first set is shot at 0600 hours. The area we are fishing is hard bottom and dotted with potential snags. This requires major concentration by Allan, who up in the wheelhouse is following plots and studying the colour sounder readouts. We proceed along a narrow track for forty-five minutes then execute a sharp turn and steam along for another forty-five-minute track. Allan adjusts our speed and the trawl length several times before giving the blast to signal haul-back.

At the end of the set Fred Marsden and Wayne Allen mend the inevitable rips before the net is shot again. Photo: Peter A. Robson.

ON BOARD THE F.V. *CALEDONIAN*

The first haul-back in our new location is greeted with cheers and big smiles. The cod end and part of the intermediate section is filled, almost to bursting, with turbot and a few rockfish. It's amazing how Allan can so accurately target his mixes. He gives the order to stand clear as winches groan and blocks creak while the cod end is inched up the stern ramp. Everyone is tense, wondering if the winches and the net will stand the strain. It is possible that the net could burst and spill the fish back to the sea. The cod end is finally aboard, squeezing under the drum and into the well with only inches to spare. The catch is split several times to empty the net. Chief explains that it is sometimes necessary to cut the net and lose some fish in order to get the remainder of the catch aboard.

The net contents are so clean that the scuppers in the well are opened and the fish run out into a hold hatch set in the deck. The few "junk" fish are thrown out through the scuppers as the fish are guided into the hold. The hard bottom has taken its toll and half a dozen tears are repaired before shooting the net again. The total is estimated at 60,000 pounds, an excellent haul. The second set comes up with about 50,000 pounds of the same mix. It is hauled aboard, the net repaired and set again. The crew begin whispering about the possibility of filling the boat today and finishing the trip early. The third haul-back nets an amazing 80,000 pounds; and due to its size, it takes half an hour to manhandle aboard. Again, the hard bottom has taken its toll, the net is badly ripped and a rib-line in the cod end has burst. The net had snagged twice on this set.

The fourth set is smaller, perhaps 25,000 pounds. The net accidentally becomes wound and jammed between the outside of the drum and its mount. It takes about twenty minutes, using cables and hooks and bits of chain, to free it up. Allan announces that the next set will be the last and is answered with cheers and "yahoos!" The weather is holding and looks good for crossing Queen Charlotte Sound tonight. Spirits are high as almost two days will have been cut from our trip.

The fifth set comes up with a satisfying 50,000. The day has netted an amazing haul—close to a quarter of a million pounds of fish. Allan grins while saying that the day's catch is the biggest haul of bottom fish he's ever had. He adds that he thinks this is the best day's work he's heard of on any local trawler.

As soon as the net is aboard, we speed up, turn, and aim across Hecate Strait. The holds are full and Vancouver is only thirty-six hours away.

Winches groan and blocks creak while the cod end full of turbot is inched up the stern ramp. Photo: Peter A. Robson.

Illustration by Angela Knight

THE *HECATE MIST'S* PHANTOM

Michael Skog

I am not the kind of guy who welcomes unexplainable visitations. I steer clear of Ouija boards, and do not do seances. I see my imagination as a liability after dark, and am basically a chicken concerning matters of the occult. That is why I've never been particularly eager to talk about the events I am about to describe. A lot of people would no doubt have welcomed the experience and made a career out of recounting it, but not me. It was a long time before I admitted it happened at all.

The beginning is always a good place to start. Colin M, a friend of mine whose name is highly regarded in industry circles, was travelling into Tofino aboard a packer in the early eighties. They were there to tow a recently purchased steel dragger back to be overhauled in Vancouver. They arrived after dark.

At the very least, the dragger *Hecate Mist* was in dire need of a paint job, for the steel hull and cabin were deeply pitted with rust. It looked like a shadow when the packer's spotlight grazed it that night. It hunched low in the water with the posture of a work boat that was used to shouldering the burden of heavy seas. It had the cadaverous look of a derelict. The packer idled up beside the derelict.

As the only deckhand, Colin's job was to jump aboard the beaten old hulk, fit it with a tow line, then untie it from the wharf. However, the instant he set foot upon the deck he recounts a feeling of anxiety laced with dread. He assured me that he has never worked so fast in all his life as when he performed his duty on its decks. The frightening sensation never entirely left him when he re-boarded his packer. Under tow, the *Hecate Mist* continued to inspire a dread akin to a solitary walk among gravestones, even in daylight.

It was possible that my friend was having fun with me and was embellishing his account to play on my overactive imagination. He is like that. Yet I caught sincerity in his tone, and there was something he said about the sensation that made it unique as a fingerprint. Colin's words came back to me in all their vividness a few summers later when I was hired as engineer aboard that very boat, but my first impression of the *Hecate Mist* was nowhere near as ominous as his. We first met in the morning of the day we were going to leave. The boat was painted a cheerful green with a white cabin, and floated squat in the water. A good sea boat, well suited for its career in dragging, but we were planning to use it as a salmon packer. Upon closer inspection, I did notice the deep rust pits in the finish, belying an eventful past. I paid little more attention to its complexion than that, as I soon began familiarizing myself with its makings and fixings.

Along with myself, there were four other crew members. The skipper was my uncle, a longtime halibut fisherman who had been urged to take better care of his back. The cook was my aunt. This was her first voyage aboard a work boat, but she has since made a long career on the sea. The deckhand, and the only person not related to me, was another greenhorn named Al. All of us were scrounged up in a hurry the day before, all of us had other plans for the summer.

Because I already had a few years of fishing experience prior to our setting out and knew how to get from point to point, and also because we had two green crew members out of four, I took the wheel by myself at night. I was pleased about this because I rarely fall asleep standing up and loved the peace and quiet of a solo wheel watch. Unfortunately, it was on this first night that the resident spook decided to pay his first visit.

For the first hour I was having a really pleasant time, with one ear listening to the VHF and the other listening to some old tape. The only problem was that my personal bilge was in desperate need of pumping. Instead of risking being caught in the toilet while ramming a deadhead, I chose to take my problem to the railing just outside the wheelhouse where I could keep a watchful eye. While I relieved myself, I glanced back along the side of the boat. There, a dark figure was staring at me. I saw it for only an instant before it pushed off the railing and vanished behind the cabin. Upon finishing my business, I went back through the galley to investigate who might be awake at two in the morning.

There was no one in the galley and no one on the back deck. I even went to look behind the two trawl drums mounted on the stern. I found no one. My thoughts began to race, reconstructing the shadow that was forming nefarious images. On my way back through the galley I reached into the cutlery drawer and pulled out a spoon for defensive purposes. I guess I figured that if a ghoul was creeping around in the middle of the night and was to attack me—wrestle the weapon out of my hand and use it against me—I would rather be stabbed to death by a spoon. You get crazy around ghosts. I waited for it to come and get me, perhaps reach in and grab my throat with its claws, if it had any. I waited there for a while. Before long, my fear subsided as I concentrated on navigating up Johnstone Straits. But the excitement was not yet over.

I was standing beside the port window in the wheelhouse an hour later when I experienced a sensation I can't begin to describe, except to say it was about the worst feeling I've ever had. Terror ripped through me like a lightning bolt. A chill came over me like I'd been flung naked against a sheet of ice. I sensed a presence just out the window, right beside me. What is more, I could see myself from its perspective—meaning, I saw it seeing me. Being the chicken shit that I am, I chose to wait until a few seconds had passed before I looked towards the window. I could no longer sense that anything was there, and when I looked, I saw nothing.

It happened again in the engine room a few nights later. The *Hecate Mist* was equipped with a champagne system that would lower the water's temperature by bubbling the tanks. As engineer, it was my job to attend to this system every four hours. I was busy down in the heart of the ship one evening, repeating a duty I had done a hundred times before, when I felt that awful sensation once again. The presence could be felt in the shaft locker, which I was standing in front of. Once again I had the very unsettling sensation of seeing myself through someone else's eyes. Then, I felt the presence begin to move towards me.

I dove out through the engine room door onto the floor of the sleeping quarters, waking up Al. As I spun to lock the door he asked me what was the matter. What was I going to say? The boogie man was after me? I don't recall what I told him—I just know it was anything but the truth.

By this time, mine weren't the only nerves that were shot. The cook, my aunt, was sensing strange things and didn't argue a bit when I suggested that the ship could be haunted. The deckhand was having bad dreams. Once he awoke us all by pounding on the roof of his bunk and screaming "Let me out! Let me out!" He thought he was trapped inside the fish hold as something stood on the lid. This same fellow accused me of making the pounding sound we sometimes heard on the side of the hull late at night. That was enough for me to suspect him of the same thing; that is, until the pounding commenced when we stood in full view of each other while the rest of the crew was snoring.

I was enormously relieved when I heard the news that we had to return to Vancouver with a load of sockeye. Up till that point we were delivering to Prince Rupert. Through the grapevine we also heard that this was to be our last trip of the season. It wasn't only because of the spook this struck me as good news. I was eighteen and had passed a whole summer without once beholding a single bikini. I was determined, upon reaching home port, to collect my paycheque and eventually find a girl in a skimpy little band-aid outfit. With this happy thought, and others like it, I passed an enjoyable last wheel turn in the early hours of the pre-dawn. But the phantom felt obliged to say one last goodbye.

Staring at the deck, lost in thought, I was again seized by the enigmatic presence. Chills surged up and down my spine. This time I sensed the entity directly in front of the wheelhouse. Realizing that it was in a different place this time made it seem more real. For a fraction of a second I feared that I might actually see this soul, but then I decided I wanted to after all, to prove that it was more than fancy. I looked up and I saw nothing, yet the chills didn't leave me. It was as if it were there to be felt, not seen.

I tried to occupy my mind with other thoughts. I'd just passed Northeast Point on Texada Island, so I decided to pinpoint my location and take a new compass heading. I looked into the radar, then turned to look at the chart. In doing so I flicked on a lamp. Its glow highlighted droplets on the window, caused by the gently falling Scotch mist. The droplets were perfectly spaced, so that you could barely fit the thickness of a dime between each one. The pattern was uniform everywhere except for the centre wheelhouse window, where there was a space where no droplets formed—the same place where I still sensed the presence. This clear space formed the exact shape of a left hand. I nearly filled my drawers.

I thought about waking up the skipper, or someone else to bear witness. I didn't, largely because I barely believed what my own eyes had seen. I reached over to the lamp, turned it off, and tried to forget about the invisible hand with the power to repel moisture. The pragmatist in me tried to explain away the phenomenon. The anchor winch was leaking hydraulic oil. The handprint that had just terrified me was probably my own. I must have accidentally left an oily handprint on the glass. With this bit of deduction, I felt so much better that I decided to confirm it by taking another look at the handprint. I turned on the light and once again saw the pattern of droplets. Only this time the pattern was perfect—there was no trace of the handprint—no trace of any presence whatsoever.

That was the last of my weird experiences aboard the *Hecate Mist*. Before closing, I should say you won't find the *Hecate Mist* in any fish boat directory because I changed the name of the actual boat out of respect for its current owner—in case he, or she, might one day wish to sell it (haunted boats are hard to sell). It might also be better for you, the reader. Some night in a strange port you may end up tied to it—and you will sleep better not knowing. 🐟

La Pérouse Bank

FISHING HAKE ABOARD THE *SOUTHERN RIDGE*

story and photos by Michael Skog

Once, BC fishermen considered Pacific hake (also known as Pacific whiting) not only a garbage fish, but a nuisance that choked trawls, seines and gillnets. Local processors had little use for it either; the flesh had a tendency to break down shortly after harvesting, leaving a pulpy mess. Yet foreign factory ships harvested it off our shores for decades. Not until the mid-seventies and the establishment of the 200-mile offshore limit did Canadians lay claim to this species. The factory ships still came for hake, but they had to share with Canadian fishermen.

The Hake Consortium was formed in 1979 to co-ordinate the partnership between BC trawlers—who prefer to call themselves draggers—and the foreign factory ships that at first did all the processing. Canadian fishermen organized their own participation in the consortium, a joint-venture project. The governments involved established the quotas, as well as the percentages of the catch to be taken by Canadian boats and by foreign boats. Priority was given to the factory ships of nations that allowed the greatest percentage of the catch to be taken by Canadians.

The *Southern Ridge.*

FISHING HAKE ABOARD THE *SOUTHERN RIDGE*

This arrangement continued, unchallenged, until the early nineties, when local processors became interested in tapping into the lucrative surimi market (a paste-like substance made from hake and used in making secondary seafood products such as artificial crab meat). In response to the shore processors' requests, the government started to phase out the joint-venture fishery in favour of a shore-based fishery.

Kirk Car, his crew, and his ship the *Southern Ridge* have been fishing hake since the season began in late May 1995, two months ago—a season that corresponds to the hake's northern migration pattern.

Until the mid-1990s, the hake fishery was a treat for tired dragger crews who viewed the opportunity to fish hake almost as a vacation from their usual dragging routine. The fish were easy to catch, and when fishing for foreign factory vessels the crews merely had to pass over trawls full of the specified amounts of hake. For crews and owners, the fishery wasn't exactly a windfall, but it did help pay the bills.

Times have changed, however. New hake products and new processing technology have opened new markets and made the fishery more lucrative for fishermen. Not only do trawlers harvest this species but seine boats now undergo gear conversions to share in the haul. More and more vessels are trying to enter the fishery every year.

The crew aboard the *Southern Ridge* planned to fish hake for as long as the plant and the season permitted, with the possible exception of fishing the valuable Adams River sockeye salmon run that passes by their area in August.

The boat is probing the waters between Bamfield and Tofino off the coast of Vancouver Island. The *Southern Ridge* is harvesting this mid-water species exclusively for a shore-based processor in Ucluelet, on the western lip of Barkley Sound. This company has an insatiable hunger for fresh product. It needs to run fish constantly through the plant because even a few hours of down-time permits bacterial growth and requires operators to sanitize the plant at a considerable cost. Feeding this mechanized giant keeps Car and his crew fishing and delivering daily.

This is no dragger's vacation. The day begins at midnight. The crew begins to move like wraiths, making ready to throw off the lines. It isn't a full night's sleep, but as much as their hectic schedule allows. A few hours stolen here and there is the norm. Ten minutes later the *Southern Ridge* is navigating through the darkness. A few lucky crew members return to their bunks to steal a little more sleep during the four-hour trip between the dock and the grounds.

Kirk Car on the bridge of the *Southern Ridge*.

Michael Skog

The morning looks as though it will be typical. In the wheelhouse, the youthful skipper is surrounded with a console holding enough electronic gadgetry to give Radio Shack a headache. At Kirk's fingertips are two plotters, two echo-sounders, one sonar, a Loran, and an autopilot—all of which are constantly in use. Mid-water trawling, says Kirk, would be impossible without modern electronics. There is still one relic present on the bridge—a paper chart lying flat on the chart table—but used mostly as a coffee cup coaster.

The image on the colour echo-sounder screen is a field of blue, which indicates that no significant biomass concentrations are under the sensors. The day before, at the same time and place, Kirk's gear was already in the water scooping up quantities of hake.

On this day, however, fishing is not so easy. By 0700 hours prospects have not changed. Occasionally a red blob works its way across the screen like a jellyfish on the tide, yet Kirk ascertains these schools to be far too small and an uneconomical way to spend the morning. When asked how many fish these blobs represent, he replies, "About 60 tons. But only a small percentage of it would ever get trapped in the net." Kirk is looking for greater concentrations of hake—he is looking for "worms."

The morning passes slowly, giving the lean, dark-haired skipper a chance to reflect on the fishery's direction. Kirk doesn't miss the joint-venture fishery. In fact, as a skipper, he prefers fishing for shore processors. While fishing for the foreign fleet, local boats had to take turns and could only proceed when the factory ships gave them the go-ahead. Canadian draggers had to fill a trawl net with an exact tonnage specified by the factory ship. What compounded the difficulty of achieving a set tonnage was the hake's schooling behaviour. During the long waits between requests for more fish, the schools of these fickle, mid-water fish would often break up and remain dispersed for hours at a time, making it more difficult for

The net sounder is lowered into the water.

trawl skippers to provide a steady supply.

Though the pressure to produce is still there, Kirk finds fewer demands while delivering to shore. He can now fish when the conditions are best, and there is no restriction on the quantity of fish he can deliver. The only stipulation is that he must unload whenever the plant needs his fish.

What lessens his burden is the offshore help he gets. During this trip, there are at least two other boats helping to feed the same processor. The three have agreed to pool their catch, so that when one boat prospers they all will. They share important scouting information, a lot of friendly banter on the VHF radio, and this morning all three boats are tirelessly searching for "worms."

FISHING HAKE ABOARD THE *SOUTHERN RIDGE*

What are these "worms" they're talking about? I learn that draggers have a lingo all their own. "Dragginese" would be the best name for the jargon industry insiders use over the radio waves. The words are English, but the speech resembles code, and serves to illustrate places or things that have no relevance to outsiders. In this language, worms are what appear on the echo-sounder when the boat travels over a large school of mid-water fish. The resulting long, red, serpentine image on the background of blue resembles, and is consequently called, a worm.

The net is set.

Evidently, Dragginese is spoken throughout the day. Early that afternoon, a gravelly metallic voice issues a message over the VHF: "Captain Kirk, are you there?"

When our skipper acknowledges, the voice continues. "Got a worm under me." Kirk then asks the person on the other end about his whereabouts and is in turn answered, "I'm at the end of the finger."

The transmitter's curly cord twists around Car as he looks out past the stern. "By the rat's ears?"

"No, near the Tom Saw yer," squawks the metallic voice. The nose of the boat turns in a southeasterly direction where the sun, flashing along the waves, no longer glints at us.

What, at first, seemed like covert dialogue between two secret agents turned out to be a straightforward conversation about a good fishing spot, but only if one understands the language. It was possible this colourful jargon was meant to be cryptic, but all these names make solid sense. On a chart of the fishing grounds, bottom features, such as the outline of a seamount, take shapes reminiscent of animals, appendages and other things. Although some ocean floor features have more than one name, they are as recognizable to draggers as land points are

Rick Towshuk lowers the doors to the correct depth.

Hauling back the net—time for oilskins.

Mike Blomly and Kevin Reid pull the cod end over the hatch.

to other fishermen. We are on our way to the "finger."

Once there, the echo-sounder immediately picks up large splotches of red. Kirk and the crew waste no time in unwinding the trawl off the drum and into the water. Once it is completely unwound, trailing the boat like a piece of limp laundry, the two large steel doors are lowered. These are attached to the mouth of the net and are meant to sink it as they are lowered on long steel cables attached to two independent winches. The heavy doors are dragged through the water at an angle, which forces them to separate and spread the net's mouth wide, so that incoming fish are funneled towards the end of the net.

Mike Blomly and Rick Towshuk are the crew members in charge of raising and lowering the doors, and consequently the net. This operation involves careful co-ordination to ensure that both doors are manoeuvred uniformly. Once Kirk establishes the depth the fish are schooled, he gives orders to Blomly and Towshuk to adjust the trawl. This is done with each

FISHING HAKE ABOARD THE *SOUTHERN RIDGE*

man standing behind a winch near the controls. The cables are marked to tell how much line is already out. Once at the correct mark, each winch is locked in place, allowing Kirk to fine-tune the depth.

Because placing the net at the correct depth is a function of cable length, weight and velocity, Kirk is able to raise and lower the net slightly by adjusting the throttle. His eyes are glued to the net sounder, which supplies him with two very important pieces of information: net depth and the quantity of fish swimming inside. His hands adjust the throttle in an attempt to maximize the capture. However, during this tow the information on the two echo-sounders does not jibe—the number of fish going into the net is far smaller than the amount that appears under the boat.

Kirk offers a theory. Perhaps the fish and the boat are travelling in the same direction. The crew seem to concur. They all remember similar situations where the catch seemed impossibly low when compared with the populations shown on the echo-sounder. I note a humility in the bright-eyed skipper, who seems more content commanding a boat of equals than drones. Consultation has a place among this group, and a decision is made to tow in the opposite direction. Blomly and Towshuk pull up the doors, minimizing the risk of the cables becoming confused as the boat turns. Off in the distance, the net rises to the surface and all aboard seem pleased; the fish are light and will be easy to work with once alongside the boat. As our direction changes the trawl is lowered again.

Not long afterwards, the schools of hake break up and fishing seems to be done for the day. The tow has lasted six hours. The net is brought to the surface and the doors are secured to the enormous stern davits on each side of the boat. The trawl net is attached to the drum and reeled in until the plump cylinder of fish brushes against the hull. The net contains an estimated 25 tons.

The procedure for getting the fish aboard is simple. Since the only

Pumping out the catch.

The crew of the *Southern Ridge*: Kevin Reid, Kirk Car, Mike Blomly and Rick Towshuk.

opening left is at the very end of the net—the cod end—it is the only part to come aboard. The cod end is pinched off by a strap and hoisted above a table meant to distribute the hake into the specified hold. The difficulty of this operation is compounded by the lumpy groundswell. As Kirk handles the controls, Blomly, Towshuk and fourth crewman Kevin Reid pull the bag of fish towards the centre of the hatches. The fish left in the water move freely in the corridor of net left over the side and reduce the load on the boom winch. Once the fish on board are dumped and the opening secured, the whole net is thrown overboard, allowing more hake to shuffle towards the cod end. To facilitate this, another strap is lifted at the end opposite to the hind orifice. Fish flush obediently along to the knotted opening so the process can be repeated. This continues till the net is empty.

As the crew clean the deck, the *Southern Ridge* heads back to Ucluelet under the same dark sky which it left twenty-four hours earlier. The morning is spent unloading. The plant grumbles and clanks in the background as it devours the catch. Busy shore workers bustle from one place to another with a secrecy of purpose. A fisheries observer shuffles in her boots down the gangway to the platform used to vacuum the catch from the hold. She writes down the significant catch information on a steno pad while the vacuum pump hisses and spews in the background. The machine yearns to be fed, and it's the *Southern Ridge's* bounden duty to do the feeding. A few hours later Kirk and his crew are heading back to the grounds, like worker bees charged with the feeding of the queen.

JIGGING AND TROLLING FOR ROCKCOD

Previous page: Sue Milligan with a string of rockfish aboard her cod fishing boat *Henry Bay*. Photo: Sue Milligan collection.

Sansum Narrows

Early photo of the Newman family at Sansum Point, late 1930s.

A LIFE IN COD
Raising a Family in Sansum Narrows

Rob Morris, photos courtesy of Mary Newman

Traversing the boiling upwellings and tide rips of Sansum Narrows, the hourglass constriction that separates Salt Spring Island from Vancouver Island, it's hard to resist peering into the swirling depths. Down in the gloom, where the currents sweep the feed along and spill it into back-eddies, is the domain of the lingcod. Over the years, many cod fishermen have plumbed the Narrows for the big lings, feeling the bottom with spreader-bar jigs double-hooked with live herring, or trolling with codskin lures.

In 1930, one such fisherman, Lauri Newman, and his wife Rosa May moved their family to Sansum Narrows to begin a life of fishing the cod. His parents had come from Finland in the early 1890s and settled at nearby Saanichton, halfway up the Saanich Peninsula.

Lauri Newman first worked as a steam engineer on tramp steamers, then moved over to steam tugs in the Strait of Georgia. Home was

A LIFE IN COD

The *Elrose* steams into Musgrave Landing, early 1940s. She was home to parents, seven children and a dog.

Lauri and Rosa May's children Mary and Dickie with the day's catch. Along with sister Clara they fished the Narrows in their 18-foot "popper" boat.

Vancouver, then Victoria. When his steam engineering career looked like a dead end, he decided to try cod fishing in Sansum Narrows. That was the beginning of the Newman family's long habitation on the edge of that turbulent, tidal passage and a long association with their bottom-dwelling quarry.

The Newman family settled by Sansum Point and the father worked his first cod boat, a steam-powered little vessel called, simply, "the little boat." Soon Lauri and his brother George built the *Elrose*, a 34-foot cod boat with a live well to keep the fish alive after being caught. The *Elrose* became home for the family—parents, seven children and a dog. Still a lover of steam when the gas boat era was well established, Lauri installed a triple expansion steam engine in the stern of the *Elrose* with the live well set amidships. This configuration avoided having to pass the propeller shaft through the well's watertight bulkheads and left room for living space forward, under her raised foredeck and wheelhouse.

On the *Elrose*, Lauri would handline in the Narrows, and when the tides were smaller, he'd range down to Sidney, and further south toward Discovery Island just off Oak Bay. The *Elrose* could carry a week's supply of wood in her stern, and there was always a willing crew of young stokers on board to fire the boiler. Only during the Second World War, when the seamless steel used in the boiler tubes became unattainable, was this dyed-in-the-wool steam man persuaded to install a Gray Marine gas engine in the *Elrose*.

Tuesday was killing day. With the *Elrose* tied up in Sidney, Father would be up at four in the morning, with the children roused soon after to help with the

Sons Bill, Dickie and Curly on the bow of their cousin Walmie's cod boat, the *Twin Oaks*. Dickie's first cod boat, *Patti Page*, formerly the *Carlisle 75*, is seen here tied to the dock.

Dickie's *Lady Jane* and brother Larry's *Northern Challenge* at the Newman family wharf in Sansum Narrows. Both salmon trollers are Frank Fredette's designs. "Uncle Frank," as the Newman children knew him, frequented the Narrows in his own troller.

The 34-foot *Rosa May*, Curly Newman's live-tanker. She was converted for salmon trolling, and has aft accommodation.

killing and cleaning of the cod held in the live well. The fish were delivered to Victoria, where they usually fetched from 6 to 12 cents a pound.

It wasn't long before Rosa May and Lauri's older children were in command of their own "popper" boat, a sturdy 18-footer powered by a small Wisconsin gas engine. They handlined, kept their catch in the boat's live well, then transferred the fish to a cedar slat live holding cage hanging off the family float. There were lots of cod to be caught, and, as Mary reminisced, it was a really good life, though her nostalgia was tempered by memories of just how cold it was in the small boat those early mornings after the cod season opened on March first.

A LIFE IN COD

The *Jaana N* is a Frank Fredette hull with a wheelhouse designed by Bill Garden. She is Bill Newman's 52-foot troller-long-liner and Roberts Creek is home port.

The Newman family together at Sansum Narrows, around the late 1960s. Back (L to R): Lauri, Rosa May, Larry, Rosie, Mary, Dickie, Curly, Bill. Front: Janie, Clara, Christa, Nellie.

The Newman boys got their own live-tankers and fished the Narrows and the waters of the lower Strait of Georgia for lingcod, eventually turning also to salmon trolling in the Strait. With the exception of the *Jaana N*, the Newman family boats were all built on the narrows—the *Elrose*, the *Twin Oaks*, the *Lady Jane*, the *Northern Challenge*, and the *Rosa May*. All the boats still have varnished wheelhouses—a Newman family tradition.

Strait of Georgia

Tor Miller's 34-foot live well cod boat, *Torhaven.* The idea of the live tank is to keep the bait alive during the trip and the catch alive until "killing day."

TOR MILLER— "SKIN ARTIST"
A Veteran Gulf Fisherman Shares His Techniques for Trolling Lingcod

story and photos by Rob Morris

Tor Miller has fished lingcod, and only lingcod, in the Strait of Georgia for fifty-one years. "I fished it all—live bait, dead bait, skins, pelks," he says about the several ways to fish the big ling. His double-ender, the *Torhaven*, is a live-tanker; its 11-foot-long fish hold can be flooded via 96 pluggable holes spaced between the ribs below the waterline. The idea of the live tank was to keep herring baitfish alive for the duration of the fishing trip and the cod alive until "killing day." According to Tor Miller, it was the Japanese fishermen who brought the live tank method over from their homeland in the 1930s.

The *Torhaven* was built in 1938 for George Springett by Hugh Rodd at his Canoe Cove yard near Sidney, BC. The only live-tanker Rodd built, she is 34 feet long with fir planking on oak ribs with 4-inch-thick bulkheads, a sturdy structure to withstand the movement of up to 11 tons of water when flooded.

Tor Miller fished in the southern part of the Strait, around the Gulf Islands. There, the cod fleet would jig for herring in Active Pass, Porlier Pass, or Sansum Narrows— each spot being more productive at different tidal cycles. With both live and dead herring as bait, he would often use a Japanese jigger, a 5- to 6-foot bar bent into a shallow curve or inverted "V" that was suspended by a line tied to its middle. This device was hung underwater, with bait attached by fishing line to both ends of the bar. ("Single hook artists" would bait the herring on one hook, but Tor preferred two hooks on each leader, one through the herring's nose, the other through the tail.) Then

TOR MILLER—"SKIN ARTIST"

he would take the boat up-tide, beyond a reef, and allow himself to drift back over the place he expected lingcod to be. With enough herring in the live tank, he could move around according to the tides for several days before needing more bait.

"I fished from Active Pass right around to Sooke," says Tor. "On Kanaka Bank, now called Beaumont Shoal, east of Victoria, or at Race Rocks or Beechey Head, you could get maybe four to five days fishing in on either side of slack [tide], during the smaller tides. You couldn't fish the middle of the big tides. But up at Active Pass, or at East Point on Saturna Island [fishing] was better on the big tides."

But, fishing around the tides with herring as bait was not Tor's favourite way to spend a day. Tor Miller is a "skin artist." His favoured method of fishing lingcod is trolling with "skins"—lingcod skins cut and wrapped around leaded hooks. He owes all he knows to Cecil Joyce, from Campbell River, "a real skineroo," and his fishing partner for many years. "With skins you can fish all the time. You don't have to keep running back for bait and playing nursemaid to a bunch of live herring," says Tor. He remembers nights that were short on sleep because he worried about the herring in his hold; sometimes the warmer water temperature in a bay could turn them belly-up.

"Fishing skins is a whole lot different than drift fishing," says Tor. "You can stem up into the rip and herring will fish on its own, but you gotta move skins over the bottom." With a 35-pound iron cannonball at the bottom of the gear, Tor is bouncing it, "feeling" the bottom all the time. For this reason he prefers iron over lead. "Iron will hang back and spring off the bottom. Lead always wants to hang straight down and stick, you're always fighting it." And although he has a sounder, it's the lining up of his landmarks that tells Tor when the bottom is coming up or going down. "You get onto it so that you don't need a sounder except maybe if you're searching out a new area."

The view down through the hatch into the live-fish hold, where wooden bungs and cloth gaskets plug the holes in the planking while *Torhaven* is tied to the dock.

A large-diameter handlining gurdy and a smaller trolling gurdy are set amidships to handle the cod gear. Both are belt-driven off the Easthope engine. Tor holds the Easthope's throttle extension, and a simple chain steering station is mounted beside him on the cabin trunk.

All the gear is set and pulled from amidships and is within easy reach on the *Torhaven*. A thirty-year-old Simplex gurdy is at the base of the mast and next to it a large-diameter Japanese-built handlining gurdy. Both are belt driven off the flywheel via a jack shaft. The Easthope's timer and throttle controls are extended by rods out the companionway and there is a steering station to starboard on the cabin trunk. The trolling line leads from the gurdy up and out through blocks on the main boom and the end of the davit which angles out at 45 degrees over the starboard side. The line angle is watched constantly and fine tuned with adjustments to the Easthope's revs. "With the davit amidships you can pivot on the gear, just to starboard, mind you, and there is lots of extra line on the gurdy so if you

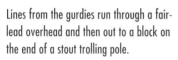

Lines from the gurdies run through a fair-lead overhead and then out to a block on the end of a stout trolling pole.

get hung up on the bottom, you have plenty of room to turn and roar back up the tide over top of the gear."

Tor's gear is very strong. "If you hang up you want to be able to stem ahead and rip the gear out of there all in one piece," he says. He adds that trolling for cod out of the stern of the boat just doesn't work as well because you lose that pivoting and tight turning ability.

The goal of tying skins was to make a lure that looks similar to a squid, a natural food source for the lingcod. Tor would always try and get his skins from a 3- to 4-pound cod. They have smoother, "fluffier" skins that fish better than the coarse skins from older fish, he says. He doesn't use dark-coloured skins, either, because they just don't fish well. He would usually change skins twice a day, for they get stretched or "tashed" out, particularly on "sticky" bottoms. The master skinner never cuts more than five tentacles on a skin and he squares the ends, rather than leaving them as points. He also bends the barb of the hook upwards, towards the head of skin—fish get stuck on them better that way.

As with all fishing operations, constant innovation and experimentation have been an ongoing process aboard the *Torhaven*. Tor found that

TOR MILLER—"SKIN ARTIST"

his own round, casted leads on the hooks fished better than the regular, oblong leads. He also tried many different colour combinations of the plastic streamers which are tied next to the skins to increase the cod's attention. Some variations did better than others.

Tor never narrowed down the lingcod's biting behaviour to a science. "One thing I could never understand—some mornings the fish would go to the top of the gear, [or upper lures], but in the afternoon the top would go dead and the bottom would load up. On a dark day they all might be climbing onto the blue skin." But Tor found a combination of colours, weights and leader lengths that worked on the *Torhaven* and he stuck with it. He tried the manufactured plastic squid instead of his homemade skins, once.

A lifetime collection of bought and hand-made jigs hangs from the cabin shelves.

A selection of Tor's cod gear. In the centre is a jointed "Ogopogo" plug; around it are several hooks with oblong leads. Two of Tor's round lead hooks are on the right. Note the added spread at the mouth of the hook.

"They seemed to work okay up around Campbell River, but I could never get them to work for me down south." He had great success when he fished with the big silver Norwegian cod jigs, or "pelks," but even these were popped into an old sock and hammered to an exact curve so they would flutter to the bottom just right.

As for knowing where to find the lingcod, Tor tells of those who will say the fish are all along the tops of the reefs. But he says: "The best fishing I ever had was on Constance Bank, away from the reefs, in 30 to 40 fathoms, where you couldn't hang up your gear if you tried, and 12 pounds was a small fish." Those were his glory days of cod fishing, but, even then, things were unpredictable. Tor never knew, from one day to the next, just what the fishing would bring.

In the early nineties, for the first time ever, the commercial lingcod fishery in the Strait of Georgia was closed. But, Tor Miller figures he has had enough years at it. "Cod fishing doesn't owe me a thing," he says. But he hopes, especially if the price for lingcod improves, that someone up the coast will be able to try his favourite method of fishing, and benefit from his knowledge as a "skin artist."

Yuculta Rapids

Sue Milligan with a 42-pound lingcod.

THE LEARNING CURVE

Sue Milligan, photos courtesy Sue Milligan collection

When I started fishing cod in 1980, there was still enough lingcod on the south coast to make fishing it worthwhile. But it was tough going, so I decided to fish live rockcod instead. One of the most appealing aspects of fishing for the live market is just that—keeping the fish alive. When the water is clear and cold, and the fishing is good, after a few days the holds look like a beautifully kept aquarium.

The quillbacks make up most of the collection, with copper rockfish and colourful kelp-greenlings sharing the holds with lingcod and some yelloweye rockfish. Greenlings are one of the few fish that can be identified as either male or female by their colour. Males have blue markings, ranging from turquoise to electric blue, and females have bright yellow to orange markings.

I've been fortunate enough to make the acquaintance of some of the people at the Vancouver Aquarium, so if an interesting fish has no commercial market, we save it for them. We bring in red Irish Lords, many different kinds of sculpins, the odd octopus, and anything else unidentifiable or unusual. A few of our larger lingcod, a wolf eel and a 35-pound halibut have all found homes behind the glass of the Aquarium.

Percy Redford, who really started me fishing cod, suggested I try the Dent, Yucultas and Arran Rapids up at Stuart Island. I fished my boat, the *Henry Bay*, around there from the spring of 1980 till the fall of 1984, with trips to Lasqueti Island, Nanaimo, Surge Narrows, Texada, Kelsey Bay and Johnstone Strait thrown in to prevent boredom. It was hard, though, to find a place that produced like the Stuart Island and Bute Inlet area. Out of approximately sixty-five trips, averaging three and a half days each, I took 112,000 pounds of rockcod, 4,000 pounds of red snapper, or yelloweye rockfish as they are now called, and 5,000 pounds of lingcod. Those years and that area paid for my boat.

It's a ten-hour run from my place in

THE LEARNING CURVE

Jervis Inlet to Stuart Island and a twelve-hour run back to Egmont, where we delivered the fish. We would run up one day, and if the tides were right, we could get into Emerald Bay just inside Barber Pass, get bait from John Davis, and be fishing by first light. By dark on the third day we usually had a load and would run all night back to Egmont to deliver. We would then clean up, go home for a day or so, and head back up to Stuart Island.

You could fish in nearly any weather there—the northwesterly wind was a bit miserable sometimes,

Sue Milligan's *Henry Bay*. "When I'm working on the boat, I sometimes wonder why I'm there instead of at home baking cookies—but I love the life."

but there was seldom any sea, which is very important when you are trying to keep your fish alive and in good condition. The water is cold and there is plenty of flow, so there was no problem with fish dying at night when we were tied up. However, the fresh water coming out of Bute Inlet presented a problem when we fished the outside of the Arrans.

When you bring fish up from colder, saltier water into fresher, warmer water such as that in Bute Inlet, more than half of them can die. So I usually fished the Bute side on the last day of the trip. The *Henry Bay's* holds are all flooded, so we carry an insulated fish box on deck to ice down the dead ones. The box holds 500 pounds, and when that's full, the trip is pretty well over.

The Rapids provided excellent fishing, but they were tough grounds to learn on. In November 1979, the first year I had the *Henry Bay*, I decided to do the last cod trip of the year along with another boat, the *Westville*. The *Westville* was towing a 24-foot herring barge, which we had anchored behind a point. However no herring showed up—only dogfish, and lots of them.

About 1800 hours, two local sport fishing guides came over in a fibreglass outboard to see how we were doing, and suggested we try over in Big Bay where they had just seen a big school of herring. My common sense told me that I shouldn't go through the Yucultas in the dark, on a full ebb, but I let myself be talked into going—"there's no problem, just stay on the Stuart Island side and you'll whistle right into Big Bay. We do it all the time."

Right. They did it all the time in a shallow-draft, high-speed boat, while I had a slow, deep-draft, half-sunk fish boat. At that point, I learned one of the lessons about listening to local knowledge: know where it's coming from and how it applies to you and your particular boat.

However, I set out and not a quarter mile past Kelsey Point the edge of a whirlpool caught the *Henry Bay*. As I held the wheel and steered her away, I felt the steering chain snap. I swung the wheel both ways, just in

case some magic happened and the chain hadn't really broken or it might have fixed itself; however that was not the case.

The herring seine was piled on the lazarette hatch so we could not get to the rudder and broken chain, even if we could have done anything. I called the *Westville*, who was behind me towing the herring barge. By the time it manoeuvred alongside and we got some lines on—in the dark and on the rapids—the tide had us all.

The herring barge was on a long tow line and it decided to make circles around us. I had no sooner said "we'd better cut that barge loose and get rid of it," when the line wrapped itself around the *Westville*'s wheel and she stopped dead. The barge took another circle route around us and decided to come straight at us. It hit the *Henry Bay*, right on the bow stem. Fortunately. The collision slowed the barge down and brought it neatly alongside. We were able to tie it up again, but only after it had done all the damage.

So there we were, the three of us tied together, going around in circles and heading straight for the unknown. I ran into the wheelhouse and looked at the compass to try and get a heading, but it had been going around in circles so many times it didn't know how or where to stop. After what seemed a very long time, but was really only minutes, we shot through Barber Pass and gradually slowed down and drifted calmly in that area that has no name, between the Arrans, the Dents and Gillard Pass.

I called the Coast Guard, and they sent out Terry Brimacombe in his 26-foot fibreglass boat to tow us into Andy Lohvinenko's old place, just east of the Arran Rapids. We spent the night there, and the next day we unloaded the herring seine and fixed my broken steering chain. I put on my diving gear to cut and unwind about a hundred feet of tow line from around the rudder, shaft and wheel of the *Westville*. After that, we decided to head up to Village Island in Knight Inlet and try to set there.

We never did get any herring that trip, and I never did master that seine; I finally sold it before it drove me crazy. I am, however, still fishing live rockcod, and enjoying life. It is one of the few fisheries where you are actually fishing. You have to catch each fish individually, and when the fishing is good, it's fun. The hours are long and your feet get tired and sometimes its boring. Other times it can be scary. When I'm working on the boat, I sometimes wonder why I'm there instead of at home baking cookies—but I love the life. 🐟

Delivery day.

LEAVING FRESHWATER BAY

a short story by Joan Skogan

Rose stayed entirely still, her back to the wall, watching the door of the bar on Howe Street until Richard walked in. Then, the never-forgotten saltwater and diesel smell of his jacket surrounded her, while a waitress in a black dress smiled at him as if she were giving a party and was glad he had come.

Rose balanced her silver bracelet upside down on the table and made its curved whale design rock from side to side. When the waitress left, she let the bracelet lie still. "Where did they find Johnny?" she asked.

Richard had turned away from her to watch the men and women

in smooth dark clothes hurrying to fill up the tables around them. Rose waited. Richard would not like answering her questions now any more than he had when they were still married. He might never tell her what she needed to know about where and how and why Johnny Kay shot himself.

"Alert Bay," Richard said. "The house by the breakwater."

The waitress stood close to him when she returned with his beer and the red wine, Rose noticed. She needed both hands to lift the wine glass to her mouth, and still it danced against her teeth. Richard was twisted around in his chair again, surveying the crowd. "Rose," his voice floated back to her, "Johnny got sick. He couldn't stay by himself in the floathouse in whatever out-of-the-way channel he wanted any more."

Rose stopped herself from saying words like, Did he use his own shotgun? just to be talking. Or asking questions like, Did he leave a note? Johnny would not have changed that much.

"If he was living in Alert Bay, what happened to his floathouse?"

"Gave it away, I guess. Same as he did with the *Marian M.* when he quit gillnetting."

"His halibut licence?"

Richard shrugged. "That was a long time ago."

"Remember when we went into Freshwater Bay to see him?"

"Johnny was never in Freshwater Bay."

Rose imagined herself lying adrift on the thin current of cool, exhaust-laden air entering the bar with each new arrival. Rolling slowly, she would be crashing softly into bodies, trays, bar stools, moving back and forth forever on an air tide.

"He had his floathouse in Baronet Pass by the old mill when you were around, Rose." Richard set down his beer so he could use his forefinger to outline the pass on the table.

Freshwater Bay was still a beautiful name, Rose assured herself. Johnny could have, he should have, tied his floathouse in there. She drank the last of her wine knowing Richard was never wrong about things like Baronet Pass and Freshwater Bay. He remembered twenty-year-old storms and the running time to everywhere on the coast. He used to forget he was married, but he remembered everything else. Rose centred her wine glass precisely on its coaster.

"What was he up to lately?" Her voice sounded sharp to her own ears, but who was more used to that than Richard? The waitress was back, smiling at him again. He ordered more drinks.

"Nothing," he said. "Living." Then, "You never used to ask about him from one year's end to the next, Rose. You hardly knew him."

She was suddenly so tired she wanted to fall forward and rest her head on the table. Impossible to explain that the coast needed to be remembered perfectly down here in the city. Everything about how it had been up there was necessary, including Johnny Kay in his floathouse, being a hermit and happy, not old and sick, not dead, for god's sake. "When did you see him last?" she asked.

"In the cafe a week ago."

Rose shook her head. No. Johnny had no use for sitting around the cafe. Never did. He would rather be whistling to deer on the beach. Firing up his speedboat to go along the creek mouths and see if the last dog salmon had gone up, or shutting down the outboard on the way back in the dark to hear the wolves howling out of the inlets.

The last time Richard brought the boat to Vancouver, Rose went to the dock to hear what was happening at home—copper rockfish on new grounds; cod prices up; Johnny Kay alongside in his speedboat at dawn the week before. Richard told her he yelled up "Jesus Christ it's cold," and his hands were too numb at first to take hold of the lines to haul himself on board. There was a bottle in his jacket to spike the coffee. On his second cup, he said Richard should get into that good cod hole further north, and marked it on the chart. He said he was certain now, since last spring, that grizzly bears sharpen their claws on the same tree for life.

Rose looked past the crowded tables around them to the floor-length curtains shutting out the street, and wondered if anyone but she ever wanted to open a window in a downtown bar. She tried to imagine Johnny in the house beside the breakwater in the village, or walking down the road to the cafe.

Richard leaned toward her. "When I saw him in the cafe, he said he needed a new hearing aid. His dog ate the first one."

Rose ignored him. Hearing aids and dogs had nothing to do with needing Johnny alive in his floathouse with the water and the beach and the trees around him the way they always were. Rose drained her glass and began another. Something old, something new. Something borrowed, something blue. Something lost. Nothing gained. The wind blows. The dogs bark. The caravan moves on. Sometimes moves on, Rose whispered into the expectant silence inside her head, some caravans sometimes move on.

She tilted her glass too soon and wine pooled on the table. Never drink this much without eating, she told herself severely. Not surprising you spill stuff. Almost nothing was surprising any more, she decided. Not surprising to be alone. Not surprising to be in Vancouver instead of on the real coast. Everything changes. Everybody knows that. The inlets, bays, channels, and islands might have altered their shapes and places and particular ways for all Rose knew, now that she was no longer there among them. She set down her glass.

Richard put a napkin over the puddle of wine. "Johnny was part Hawaiian," he said. "His mother told him when he was a kid."

Rose reminded herself that Johnny had always been like Richard. Part Indian. Part something from Europe. Now, Richard was telling her Johnny Kay was somehow Hawaiian, too. She folded her hands together and sat up straight.

"Hawaiian sailors were making pineapple Indians around Alert Bay a hundred years ago." Richard looked at Rose. She managed a nod. "Johnny said his mother used to talk about taking him to Hawaii. She drowned right out here in front." He gestured widely toward the entrance to the bar.

Rose clenched her hands together, and looked at the door to Howe Street. "Not here," Richard said impatiently, "in Johnstone Strait. In front of Alert Bay. She hooked a big halibut when she was by herself in her canoe. The line tangled around her leg and pulled her in."

Richard talked into his beer so that Rose had to lean forward to hear him. "Johnny said he was tired, Rose. He said all he wanted to do was get to Hawaii."

Rose swallowed the last of her wine and saw Hawaii perfectly. Palm trees and white sand in the sun. Warm, sweet air almost tasting of flowers. The sea just easy all the time, not grey and mean, the way it is here so much, only gently rocking, like a mother. Never a sign of nets or lines or any other work gear. No storms. No need to push yourself to leave safe harbours and quiet, unchanging, heavenly blue seas.

On the street, she stood unsteady beside Richard in the rain while he waited for a cab to take him to False Creek. He looked up at the night sky. "Supposed to blow southeast tomorrow." The taxi honked and he was gone.

Rose asked her feet to keep her standing, watching after him. She decided not to name the places up the coast where it would matter when the wind changed tomorrow, and to wonder only once if Johnny Kay had ever wanted to be married. She walked with her arms slightly lifted to balance herself, noticing that the rain and the wine were softening the edges of the buildings and cars and lights, changing them into shapes that seemed new and already known at the same time, she thought, making the city easier to navigate. 🐟

DIVE
FISHERIES

Previous page: Ron DeBoer on the bottom. He's holding on to the rim of his pick bag and in his right hand is one of his custom-made aluminum urchin rakes. Photo: John Houston.

OCTOPUS DIVERS
Hunting the Giants

story and photos by Rob Morris

Lee Smith and Al Crow consider themselves the second generation of octopus divers on this coast. They were preceded by a handful of pioneers: Ray Linden—considered the "granddaddy" in the fishery—Frank and John McGuire, and Rod Palm. Crow apprenticed with Ray Linden as a dive tender, and now Smith, who started off as a tender for Crow, has in turn gained from his knowledge as they fish together.

Al Crow and Lee Smith fish the giant Pacific octopus, *Octopus dofleini*, the largest of three octopus species found in BC waters. Since its beginnings, the octopus fishery has primarily supplied bait to the halibut fishery. "Octopus has long been a preferred bait for many halibut fishermen, particularly those who fish up north," says Al. "Octopus is a natural prey species of the halibut; it's firm and stays on the hook while setting the gear and will often stay on the hook for a second set."

According to Lee there is also a small food market for octopus—primarily for customers of Mediterranean origin. However, as of 1991, local fish markets were requiring a maximum of 100 pounds a week. Generally, the food market for octopus wants a much smaller animal than the bait market, usually no more than 15 pounds. "We're targeting the larger animals; that's what the fishermen want," says Lee. "This works out well from the 'conservation of the fishery' perspective, since we're not taking the small animals before they've had a chance to reach reproductive size."

A lot of individual effort goes into the development of bait markets. And this is on top of a lot of years developing and maintaining a "paper route" of regularly producing dens.

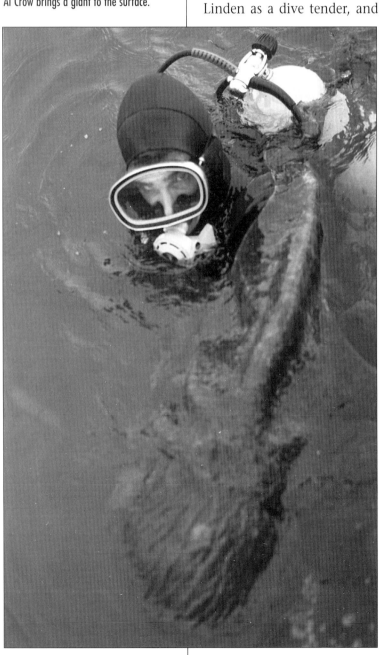

Al Crow brings a giant to the surface.

OCTOPUS DIVERS

Live octopus on the deck of Al Crow's dive boat *Outlaw.*

Octopus apparently live only about three to four years, and reproduce in the last year or so of their life cycle. Young octopus gain poundage very quickly. According to Al, a two-year-old, 5- to 8-pound animal will grow to a 40- to 70-pound breeding adult within a year. Males can reach 100 pounds; a large female is about 50 pounds. "Taking the small animals is pointless," says Al. "We're cutting our own throats if we take them before they've had a chance to grow and reproduce."

Research is showing that breeding and recruitment of octopus populations on the west coast of Vancouver Island takes place in cold, well-mixed waters all year round, with breeding peaks in spring and fall. Females will choose a suitable den, and then, it appears, a chemosensory mechanism attracts males to the den site. Breeding can last several days. The third right arm of the male enlarges and modifies as a copulatory organ, called a hectocotylus. The male charges the hectocotylus with a tube of spermatozoa which he deposits up under the female's mantle skirt, near an oviduct. After fertilization has occurred, the female uses a mucous secretion to attach her eggs to the ceiling of her den. She will deposit up to 80,000 eggs in strands of 150 to 200 eggs. Once the eggs are in place, the female continually aerates them with jets of water from her siphon and a fanning motion of her tentacles.

From the time of fertilization the female virtually stops eating and barricades the front of her den with rocks. Three to six months later, depending on water conditions, the hatchlings, maybe a sixteenth of an inch long, struggle out of their egg capsules and head for the surface, where they start to feed on plankton. Only two or three will make it to adult size. The female will mate only once in her life, the male several times. Both become senescent and die within nine months of mating.

"If you see a barricaded den with older feeding litter outside, you know it's a female on eggs," points out Al. "If in doubt, you carefully reach inside and feel the egg strands."

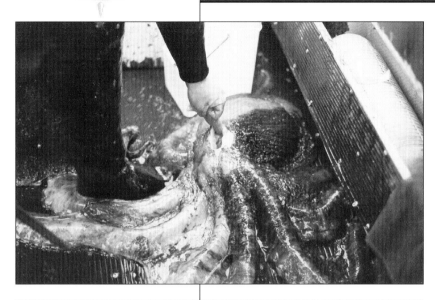

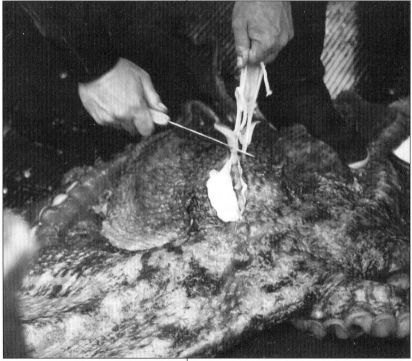

Removing the brain: this is usually done underwater to immobilize the animal.

"Freecrawlers" is the term given to octopus that are outside dens, roaming the bottom. Their numbers increase at night, when most feeding takes place, and in the spring when males are searching out females. During the day, the vast majority of animals are in their dens. Larger octopus occupy the larger dens and will evict smaller animals, often eating them while they're at it.

Aside from this tendency towards cannibalism, the octopus has its diet preference—crabs and scallops top the list, followed by various other mollusks and crustaceans and the odd ambushed fish. Al and Lee have seen the remains of cormorants outside dens, probably snagged while diving to forage among the rocks. Recently, they found the skin of a harbour seal pup in a litter pile, or midden. Animal shells and skins are left virtually intact after an octopus feeds. It will first immobilize its victim by secreting a debilitating enzyme, bite with the parrot-beak located inside the mouth opening, then suck out the innards of its prey.

Octopus harvest gear is not complex. A fisherman needs a fairly rapid vessel to search and maintain dens over a large area, plus the diving gear: a drysuit, mesh bag to bring the octopus to the surface, squirt bottle of bleach and a knife to dispatch the octopus when captured.

The bleach used is calcium-based, like the swimming pool variety. The diver squirts a small amount into the den to drive the octopus out. "Too much bleach will linger in the den and affect its returnability," notes Al. "Attempts to harvest or disturb females on eggs are definite no-nos." Al and Lee have both recently been fishing off Al's 18-foot vessel, *Outlaw*, but each diver has his own run of dens to dive on while his partner tends and dresses the catch for him. More often, though, they are running their own vessels.

An octopus will fight back if grabbed by a diver, but really all it wants to do is get away. So, according to Al, all you need do is relax your grip and the octopus will do likewise with its suckers. If totally released it

OCTOPUS DIVERS

will shoot off, propelled by water taken into the mantle and jetted from its siphon, usually ejecting a cloud of ink as a smoke screen. Some claim the ink cloud takes the shape of an octopus to fool predators. Lee has known the escaping animal to quickly change direction behind its cloud as an evasive action.

A captured octopus is dispatched on the bottom by cutting out a portion of the forebrain—both for humane reasons and because it's easier to put the animal in the dive bag. The animals are dressed up on deck. The mantle is turned inside out, the two gills, gut sac, ink sac, reproductive organs and beak are cut away, and the octopus is put in a plastic bag, whole, for freezing and delivery to the fisherman.

Frank McGuire represents the Pacific Octopus Divers Association on the DFO's dive sectorial committee. He would like to see the limiting of licences to avoid the over-exploitation of the species which is a prime candidate for overfishing. Another conservation option being pursued is the rotation of closures to heavily fished areas, such as the lower Gulf Islands, where 30 percent of the coastwide octopus landings are recorded.

Fisheries statistics from 1989 show divers taking by far the greatest poundage, with groundfish and shrimp trawls and traps, mostly prawn, accounting for the rest.

The BC fishery for the giant Pacific octopus remains a small, labour-intensive one. A dedicated and exclusive "second generation" of divers such as Al Crow and Lee Smith continues to work on a "den-by-den" basis with their elusive molluscan quarry. ⚓

The third right arm of the male enlarges and modifies as a copulatory organ, called a hectocotylus.

Lee Smith and Al Crow with a pair of fresh octopus. Freezer bags are just visible behind them in the stern of the boat.

ON BOARD THE *ROBIN D*
Harvesting Red Sea Urchins in Johnstone Strait

Rob Morris

Johnstone Strait

Red sea urchins. Photo: Rick Harbo.

Sunday, January 20, 1991

0610 hours. The *Robin D* pulls away from the Kelsey Bay government dock and heads down Johnstone Strait to the Chatham Point area and the first day of a four-day red sea urchin opening. Skipper Ron DeBoer is heading back to the spot where he and his crew left off the previous week. "We picked twenty totes on the last day, two boatloads, set a record for ourselves! With the long trip out this morning, we'll probably get at least a load, but hopefully we'll find something that we can jump on early tomorrow morning," he says.

ON BOARD THE *ROBIN D*

Travelling with the *Robin D* are the *Leading Lady*, skippered by Earl Hentge, and the *Sea Witch*, skippered by Mike Miles. Ron idles just outside the harbour, waiting for the *Leading Lady*. She's got a frozen tie-up line, but Earl's voice soon crackles over the VHF. "We're behind you, Ron. Had to use the coffee water to unfreeze my stern line!"

0650 hours. Breakfast is served up by Sam LeBlanc. He is the second diver on board with Ron and has recently graduated from seafood harvester training school, having earned the Workers' Compensation Board ticket required for this work. The third crew member is tenderman Steve Ellingson, deckhand on trollers and halibut boats for nine years but now, "trying out some different fisheries."

The *Robin D* is owned by Ron and his partner John Houston under the name Adanac Ocean Services Ltd. Both are from Nanaimo and together they have thirteen years of experience diving for red and green urchins, sea cucumbers and geoducks. Their vessel is a 35-foot US Navy surplus landing craft built in 1956 of 2-inch steel-clad fibreglass and powered by a 671 GM diesel with a straight shaft. Nicknamed the *Rockin' Robin* by the crew, Ron admits, "She's a scary boat in rough water—she's only got an 11-foot beam and the plywood cabin put on by the former owner added a lot of top weight." Plans for a 35- by 13-foot vessel are underway, but otherwise the *Robin D* is a pretty ideal dive platform. She has good carrying capacity, a bow ramp that lowers to water level for easy diver access, and a propeller that's protected in a tunnel formed by the chines and skeg, enabling Ron to venture into some rock piles where others wouldn't dare to go.

Steve Ellingson helps diver Sam LeBlanc aboard the *Robin D*. Photo: Rob Morris.

Right now the *Robin D* has only a red urchin tab. Ron and John were just putting the boat together, and had their cucumber and red and green urchin tabs when limited licensing on these invertebrate fisheries was introduced. They didn't have the required poundage to get their green urchin and sea cuke tabs and had to really scramble to get their red urchin poundage. With only one licence for now, the *Robin D* will be fishing red urchins all year. "We managed to do that last year," says Ron. "The urchins spawn later as you go north, so we were able to work the central coast up into June before our buyer told us the quality was falling off. In August we were back at Goose Island, the central coast through September and October, down in McNeill in November, Sidney in December, and now back

Diver Sam LeBlanc guides a deck bag over the hatch of the packer *Rolano*. Visible at left are the divers' air hoses hanging on the front of the *Robin D*'s wheelhouse. Photo: Rob Morris.

up here. We haven't yet had any problem selling product, so as long as we can find quality roe, we should be able to keep everyone working. The three divers rotate shifts, two weeks on/one off and Steve takes time out when he needs it."

0900 hours. The *Robin D* arrives at last week's dive site. Sam is now in his drysuit. He and Steve take up positions on the lowered ramp, looking down through the water for urchin concentrations while Ron, one eye on the sounder, manoeuvres *Robin D*. If the water is clear, a visual survey can be done from the ramp, usually with one diver on the ladder with his head in the water. Poor visibility necessitates dropping the hook and putting the diver over to search, then moving the boat again—"braille diving"—it takes a lot of energy. "The days you work the hardest are the days you make the least amount of money!" Ron laughs.

This survey reveals enough urchins to fill the boat, but the ebbing current is fast, and it is neither safe nor efficient for divers to be working in more than 3 knots of current. Three hundred pounds of urchins go into the picking bag before the diver walks it over to the haul-up line dropped from *Robin D* by the tenderman. Lift bags made from truck inner tubes and attached to the picking bag bridles are inflated, which makes the pick bag neutrally buoyant and less likely to snag the bottom, but a strong current can make the trip back to the line a real struggle. Occasionally, the divers have to wedge bags between rocks to be retrieved when the tide slackens.

Big tides are a major limiting factor on harvesting in inside waters. "Lots of days we get no product on the boat until we can get in the water at slack tide," says Ron. "If there are two slacks, all the better. North of the

ON BOARD THE *ROBIN D*

Island, tide isn't a consideration. Up there it's the groundswell that can knock you back and forth!"

The crew decides to return to this spot after noon when the tide has switched. "The product here was primo last week," says Ron. "Although the shells were a little thin," he adds. "In strong tide the urchins will hang on, and if you try and pick them with thin shells, their bottom can rip out. Our buyer doesn't like getting busted product, but as long as it is kept to a minimum and the quality stays up, we'll try here again."

A call to the *Leading Lady* reveals similar tidal conditions nearby. "It's screamin' through here," says Earl. He and his crew are in a back-eddy, so they decide to work the spot and "hope the current goes away!"

Ron heads the *Robin D* around the back side of an adjacent island to "look for product." "We look for a combination of rock, broadleaf kelp and current as usual indicators of urchin areas, but we've found concentrations on sandy bottoms with no rock or kelp anywhere in sight," he says. He noses into likely spots and Sam and Steve use hand signals to indicate upcoming bottom, rocks and the presence of urchins—one finger for so-so numbers, two for pickable quantities and three for a hot spot. "It could be one of those days when we do a lot of looking!" Ron says.

1035 hours. In the lee of an islet, two fingers go up. The hook goes over the side and after some help from Steve with his backpack and air supply, Sam descends for a survey. In ten minutes he is back on the ladder announcing, "There's product here." He goes back down to move the anchor closer to the concentration and starts picking while Ron suits up.

Each diver wears a drysuit, 90 pounds of lead ballast, and a backpack which holds an air hose connection and a pony bottle with ten minutes of emergency air. Ron and Sam are connected to the air compressor on board by 250 feet of strong 3/8-inch air hose taped to a 3/4-inch safety line.

Once the pick bag has been sent back down to the diver, Steve transfers the urchins into deck bags suspended from a beam spanning the bulwarks. Photo: Rob Morris.

Each of them can theoretically work close to a 500-foot-diameter circle around the boat.

Ron wears a pair of steel-toed gumboots over his dive boots as protection from the urchin spines. Ankle weights keep the divers upright as they walk along the bottom, scooping urchins into the pick bags with an aluminum rake. The rake fits around the forearm and has two prongs 4 inches apart, enabling the diver to use it as a gauge to measure minimum harvestable test, or shell, size. Ron designed it himself and it's a great improvement over the chopped-off garden rakes they used to use. On the bottom, the divers stagger deliveries of their pick bags to the downline with the tender sending empty bags back down. The compressor, located in the engine compartment on the *Robin D*, kicks out at 160 psi and back in at 120 psi. "We always try and dive at less than 40 feet," says Ron. "It gives us maximum bottom time without decompression. We can do 200 minutes comfortably in the 30- to 40-foot range, and if it's a good spot, we can fill the boat in that time."

A diver's backpack and rake. Photo: Rob Morris.

Steve operates the hydraulics up top, winching the pick bags aboard and dumping the urchins on deck. He shovels the urchins into deck bags suspended from a beam which spans the bulwarks and can be moved forward as the bags are filled. A deck bag holds 500 to 600 pounds and Steve can squeeze up to eighteen on deck.

1115 hours. The first pick bag is aboard. "They'll come every twenty minutes or so, but if picking is really hot, they're coming every eight minutes and I'm really flying," says Steve. He keeps his eye on the locations of the two divers, their hoses and their pressure gauges mounted on the front of the wheelhouse. The new bag system is a vast improvement over what they used in the past—the tenderman dumping the pick bags into plastic crates or cages, redistributing the urchins by hand, all the time shifting cages around the deck to make room. At the dock, the cages had to be carted by hand up the ramp to be loaded on the truck, very labour-intensive. Ron's comment about all this is that "we're working smarter rather than harder." The new rake, the inner tube floats on the pick bags, and the deck bags have all been worked into the *Robin D's* system over the last year.

1235 hours. Ron is on the ladder wanting to know if the packer *Rolano* has called on the VHF. If the *Robin D* can offload soon, the deck will be clear for the afternoon dive. Ron tells Steve he's going to move the hook closer to where Sam is picking, then he heads back down.

1315 hours. Sam is up. "It's pretty scratchy—not a lot of product." But he's made a friend down below: "I fed some roe to a little octopus—he wouldn't let go of my hand!"

1320 hours. Ralph Hull, skipper of the *Rolano,* is on the VHF. *Rolano* shows a few minutes later and drops her anchor a couple of hundred yards off. More economical than running back and forth, the three dive boats are paying Ralph a flat rate to collect their product on the grounds each day and run it to Kelsey Bay each evening where it's picked up by the buyer.

1330 hours. Ron climbs the ladder to the deck. He was into a good spot where the roe quality was very good, so he and Steve decide to file this spot away for another time when they need a place to get out of the tide.

1340 hours. *Robin D* ties up to the *Rolano* and the deck bags are off-loaded into totes. Each bag fills a tote. *Leading Lady* rafts up on the other side of *Rolano*, waiting to offload. The current is "brutal" where they've been picking, but they've done well. Sam has constructed pizzas out of nowhere. He says he has taken charge in the galley, his rationale being, "I want to eat and these guys ain't going to cook as well as me!" His advice to Ron is, "Better eat that pizza—it's going to be a marathon this afternoon!"

1435 hours. Ron and Sam are back in the water. The tide is still strong but down from the morning, and as it's a small flood, they should be able to pick through to dusk.

1700 hours. Eight deck bags on board. *Sea Witch* and *Leading Lady* are both pulling the hook to head over to the *Rolano*.

1745 hours. With Ron and Sam both up, Steve coils the air lines, pulls the anchor and *Robin D* turns toward the *Rolano*. Sam is contemplating cajun steaks for dinner, so Ron's claim that the *Robin D* eats the best of all the boats is apparently justified!

The final count from Ralph Hull is sixteen totes or, at 500 pounds per tote at the plant after water loss, about 8,000 pounds. "We can do better," says Ron, "but it's not bad for the first day out." The *Robin D* moves over in the darkness to where the *Leading Lady* and *Sea Witch* are already anchored up for the night. 🐟

Telegraph Cove

Keith Ware preparing to dive. Keith is using a full face mask underwater communication system.

TENDER MERCIES
Urchin Divers at Work

story and photos by Ian Douglas

Breaking pre-dawn ice by the Port Hardy dock chills the soul and makes two layers of polypropylene seem hardly enough for a day underwater. Before donning the drysuit, a final pump of the bilge while pondering the merits of relief zippers. Black rubber gloves balloon with a hiss, squeak when twist-checked. They'll pop before day's end—might as well start off with dry hands, though. Divers pull on "poly" liners, then cotton liners and lastly the extra large Marigolds. A final snap, cuffs drawn tight over the wrist seals bulked up by plastic pipe. Careful turns of black electrical tape complete the diver's dressing ritual. "See how I scrunch the end, leave it hanging? That's real important when you want to get out of your suit in a hurry." The mental checklist starts: Backpack hooked up, compressor on, got air, weightbelt right—bailing out and dropping a belt on an air hose is every diver's nightmare—computer, knife. He climbs onto the ladder, nods to the tender as he is handed the rake. It's another red urchin sunrise.

Before this winter, I normally avoided sea urchins. Especially after hearing a tale in a Cabo San Lucas bar, about a tourist who'd been washed backwards by a rogue wave while snorkelling off the coast. After landing on his butt in a patch of "devil's pincushions," the gringo traded his time in Margaritaville for a guest spot in the emergency room. But when a job as dive tender on an urchin boat came up, curiosity (and a ballistic Visa bill) gave me the opportunity to learn more than I ever wanted to about harvesting "sea porcupines."

Boat prep time. I watch John Fraser's hands fly as he lays in an eye splice on a net bag. The lobster fisherman from Pleasant Bay, a small village on Cape Breton Island, tells me it's the first time he's ever been paid for doing this; "Mostly we fix the gear in the netshed, wintertime, when the Gulf is all frozen up." Growing up, he had gone out trawling on his dad's boat, and presently, he fishes his own Cape Ann-style boat with 300 lobster pots. "It's a good job," he says, "up at five, setting in anywhere from 30 fathoms to right in amongst the rocks later in the season." Coming out west this fall with some pals, he went out on a seiner: "Interesting work, but didn't make much." He is heading back east for Christmas as soon as the net bags are squared away.

After compressor filters and engine filters are changed, the *Ragu*, a 40-foot aluminum dive boat owned by Eric Gant and skippered by Keith Ware, fuels up, and with Gary Gagnon aboard as deckhand, runs up to fish the opening off Telegraph Cove, an area on the north end of Vancouver Island.

A tender's job is pretty basic—look after the divers, load the product and keep things running. On a boat with a compressor and air lines, the tender ensures that each diver is suited up, hoses are paid out without kinks and the compressor keeps on sending down life-sustaining air. A close eye on the gauges is essential. Listening to the motor blasting away is not enough, as belts can slip off. Before the divers hit the water, a downline with pick bags—each capable of holding about 200 pounds of urchins—and a float to keep the bags neutrally buoyant (often an old buoy, with a hole cut so that the diver can fill it with air) are dropped to the bottom.

Red urchins aboard the *Ragu*.

Entering the water off the ladder, the divers grab a line running around the hull at the waterline and then descend the downline. Strange commute to work, but once they are on the bottom, they are down for hours, unless the current picks up too much, they run out of product, or their back teeth start floating.

Keith, lead diver on the *Ragu*, is using a full face mask underwater communication system. A mike and earphone is powered by a 9-volt battery and the topside SSB unit runs on a rechargeable 12-volt gel-cell. As long as the transducer cable is in the water (easy to forget sometimes, as it must be pulled each time before we move), Keith reports he can hear the tender well, but it is often hard to understand the diver. At one time, the tender could only hear the diver through an outside speaker when he was out working on the deck, but a new system, consisting of mobile headphones and boom mike, should allow constant communication, once all the bugs are worked out. Normal dive communication via tugs on the downline or air hose is used for the other diver. There have been some problems with the batteries in the diver's "com device" running down, but in an emergency, such as a dragging anchor, the tender can always haul in the air hose to bring the diver back to the boat.

Snagging an air hose on the downline is always a hazard when product is being lifted to the surface (especially if a carabineer, or clip, is

Tender Gary Gagnon monitors the air line from the *Ragu.*

Dive boats under tow.

used on the end of the downline, rather than a plain ring). Causing a diver to "do a Spiderman" on the bottom of the boat always livens up a diver's day. Kevin Paiuk, the other diver on the *Ragu*, told me a tale of a near downline disaster that happened to him on another boat. Working a drop-off, he started climbing the downline and was almost at the surface when it suddenly went slack. Tumbling into free-fall, Kevin remembers thinking "this guy's going to kill me," all the while trying to climb faster, but still dropping. At 66 feet, the compressor was having a hard time delivering enough air. "I saw a white anemone on a ledge and suddenly it had ten holes in it. I was hanging on, trying to figure out what to do." The downline was hanging just out of reach. Finally he grabbed it, and started up it as if a giant octopus was right behind him. "I'm sorry, it slipped," was all the tender could mumble when Kevin finally crawled aboard.

Steve, a diver from Port McNeill, tells a tale of hitching home in his drysuit. Diving on scuba off Haddington Island, the anchor dragged, and the wind blew away his boat as his tender tried everything to start the motor. He dropped his bag and belt, swam for an hour and a half, and finally stumbled ashore on Malcolm Island. He found his way up to the road with his dive gear and started hitching. The first truck along stopped, asking if he was the diver guy the Coast Guard was looking for. The driver called the Coast Guard on the VHF in his truck and Steve was back at home before his boat was towed into port.

Among the urchin fleet, some divers use rubber suits while others swear neoprene is the way to go. "I hated diving at first," says Kevin, "I was always cold in the rubber suits, but in the neoprene, once I start working, I'm warm." What with inner-tube chaps to protect legs and knees, rubber workboots, heavy weights, and the occasional bursting bladder, it is not hard to see how diver burn-out occurs.

TENDER MERCIES

Fishing is fishing, whether you are after urchins or salmon. If you are broken down, you are not making money. Most of the fleet have small gas engines running a compressor along with hydraulic fittings; for these, the salt air and salt water must be one of the worst environments. Many boats have spares, but just storing them in a marine environment means it's a battle to get them started. Jury-rigged is the name of the game—anything to get the product on board and get home. One day, at the same moment the pull-cord on the 11-horsepower engine running the hydraulics on the *Ragu* broke, the tide swung and the current started picking up. When I told Keith of the situation, he told me to send down some more pick bags so they could keep busy while I worked on repairs. While rigging up an interim pull-cord, my second pull brought the engine roaring to life as it shredded the new cord. At this moment, the boat swung as though the anchor line had parted, but it was only a false alarm, merely the current getting stronger than the gusting wind. With the remains of the cord flaying itself to death and the current gushing by, we were able to lift the bags, haul the anchor and get the hell out of there just in time.

To land the product you have to find it first. On the *Ragu*, a diver holds onto the ladder while surveying. It is a lot like flying—soaring along searching the bottom for concentrations of urchins. Seals resting in the shallows are often spooked by the sudden appearance of the boat, which will scout right in among the rocks. Every crew dreams of finding a glory hole where the divers can load up quickly. Since urchins feed on kelp, some of the best spots are where storms deposit blown kelp. Ideally, the divers hope to work in less than 30 feet, which often means anchoring in precarious situations, relying on a stern anchor to keep the boat from swinging onto the rocks, a hairy proposition when winter squalls can blow up in no time.

It is a crazy way to make a buck. Perhaps that is why, in an overheated hotel room with dive gear cascading from the curtain rails and a tower of pizza boxes slumped beside the TV, a diver might explain why he only buys the one-year personal commercial licences instead of the five-year ones. After a slug of beer, he might swear that each season is his last, "never again," he says. As for me, I have to check that Visa bill. 🐟

Kevin Paiuk with a pick ring for a double-ended bag. Each diver swims with a double-ended bag held open at one end around an aluminum ring.

Deep Bay

Dive tender Jeff Austin spills a bag of geoduck on the deck of the *Hideaway II.*

GEODUCKERS
Happy as Clams

story and photos by Michael Skog

The *Hideaway II*, a 38-foot gillnetter-turned-dive-boat, was moored among a floating labyrinth of hulls and masts at the government wharf at Deep Bay on the east coast of Vancouver Island. This small community within the bosom of the Strait of Georgia was the temporary host to an equally small fleet of geoduck fishermen who were as unique as the clams they fished.

Aboard the *Hideaway* three crew members—boat operator/head diver Dave Thomas, second diver Lonney Tewkesbury, and dive tender Jeff Austin—were already warming the engine and oil stove. Condensation had frozen into a sheet of January ice covering the deck and the crew didn't seem to be in a big hurry. Other dive boats, also hunting for geoducks, glided slowly past us—their crews offering salutatory nods—on their way to the grounds.

Much of the competitive tension was bred out of the fleet when management of this fishery was changed to a quota-based system several years ago. Now the licence holders catch their quotas at their leisure and co-operate. They all belong to the Underwater Harvesters Association, which has a lot of input regarding the management of the fishery. They also volunteer their own time and money towards regulating the fishery—a source of pride among geoduckers.

As the crew threw off the lines and eased away from the wharf, I noticed a strange totem held to the mast by a noose: a doll, mutated by fire and corrosion, holding a trident in its hand and riding a rubber duck (plastic actually). Jeff told me this ghoulish talisman was their "Baby Neptune" and served as the boat's mascot—nothing to worry about. Over the years, rival geoduckers had defiled the lucky charm by removing its original head and replacing it with

GEODUCKERS

another that sprouted devil's horns. This head was also replaced, but none of the crew can remember if the original was put back on, or if the doll now sports an entirely new head. Didn't matter, it was an old joke now with no offence ever taken.

The cabin filled with metallic voices crowing over the VHF. The nice weather—blue sky and calm sea—seemed to have generated a lot of rambunctiousness throughout the fleet, and another geoducker began to accuse our boat and its crew of shadowing them around the grounds. Dave seemed horrified that I might hear this and perhaps write it down, he straightened me out by saying such verbal abuse was to be expected. When a boat discovers clam-rich bottom, the other boats are naturally curious to find the location—the same principles that apply in all other fisheries. The targeted boat then has the privilege of calling the other boats "claim jumpers."

Unfortunately for Dave and his crew, yesterday's highliner was a friend of theirs who was especially talented at being a pest. The two divers say this playful rivalry runs amok throughout the entire geoduck fishery. And, since finding the right spot to start is the first decision of the day, it is also when the boat is most vulnerable, and gullible.

Before the days of satellites and global positioning systems, boats would thwart claim jumpers by placing their buoys (used to mark a spot worth returning to) in bogus positions such as over shoals and gravel beaches that had no chance of yielding geoducks, thus "psyching out" opportunistic rivals. However, the modern geoducker relies on a more complex form of deception. In the animal kingdom there are species of birds who fake a broken wing to lead predators away from their nest; geoduckers subscribe to this school of thought when they make a show of preparing to dive, then waiting until the following boat sets its anchor and prepares in earnest. Once the other boat is committed, the first boat buggers off to more fertile fields—all in good fun.

"Baby Neptune."

Lonney Tewkesbury assists Dave Thomas into his dive suit.

Once Dave chose a spot, the rest of the crew slipped into their preparatory routines as if silently following a well-rehearsed script. Lonney dropped the anchor and Jeff uncoiled the large water hose, the pressure from which is used as a digging tool that loosens the sand around the clam and allows for a gentle extraction. Dave would be the first to dive, so Lonney helped him get into his drysuit and make sure it was watertight. Out on deck, the young skipper strapped on a weight belt that would

A geoducker descends, armed with his stinger and harvesting bag.

ensure he would not float above the sandy bottom. Jeff paid out the air hose and patiently waited till Dave was ready to put on the breathing apparatus—divers are fed surface-supplied air, forced through an air compressor that is always closely monitored. Attached to Dave's breathing apparatus was a "pony bottle" with twenty minutes of extra air, "for emergencies" said Lonney. As good-humored as they were, the crew betrayed a sobriety when it came to their equipment and safety procedures, a reminder that their occupation was often fraught with danger. (They, of course, are compensated for the risks they take. Every individual is paid a percentage of the overall take, although the amount differs from crew member to crew member and from boat to boat.)

Finally Dave stepped down the ladder with the childlike tottering of an astronaut and vanished through the water. Steel-green and as clear as it was, the water still didn't give us a view of the diver working on the bottom, 29 feet below. The surface disturbance from bubbles, along with the curvature of the dive hose, were the only signs that indicated the diver's location. At first it seemed that once the diver was down he was totally incommunicado; Dave may as well have been on another planet. And when the two other crew members explained the terrain on the bottom—bland in colour with wave and craterlike features—it summoned images of lunar landscapes. One must feel extremely isolated and vulnerable in such an environment. He was down less than five minutes before there was a series of tugs on the rope that waited on the bottom.

"Four tugs, he's coming back up." Jeff wound the rope around a hydraulic capstan and pulled the diver back to the surface. This was just one sign of many in a physical vocabulary shared between a diver and his team. One tug would have meant to stop all surface to bottom movement; two—to give more slack; three—to bring up the clam bag. But four tugs this early in the dive meant there was a problem. As his mask broke the surface, he spit out his mouthpiece and an embarrassed smile spread across his face. "There's nothing down there." Apparently, the bottom had been picked clean a short time ago and there was nothing but holes made by the last diver's pressure nozzle, or "stinger." Did they fall for the old bait-and-bugger-off routine? As if they would ever tell me.

We hoisted the anchor and dragged the hoses a short distance to a shallower spot. Within ten minutes the man on the bottom tugged sharply on the rope, indicating, in diver lingo, there was a bag to be emptied. The white-meshed bag came up taut as it broke the surface with the weight of approximately 150 pounds of mollusks. Jeff eased the dangling bag over a blue, impact-absorbing foam mat used to keep the catch in pristine condition. He lowered it till the lasso ring on the bottom of the bag grazed the deck, allowing him to safely loosen the release knot.

In slow motion, enormous geoducks spilled over the mat. They shared little in the way of size and shape, for some were longer than the average arm and hovered around the 9-pound mark, and some were short and light. They ranged in colour from the most valuable white-fleshed, light-grey-shelled clams to the less valuable silt-coloured clams with dark grey to black shells. Yet most of the geoducks fell between these colour extremes, in a textbook example of genetic diversity.

Jeff packed the catch into plastic "cages"—industry-standard containers. The day was going well for the dive tender, for calm water is a two-fold blessing. Not only was there no pitching to make the human passengers uncomfortable, the absence of a swell prevented the geoducks—which are extremely delicate and prone to shell breakage—from rolling around. The tall dive tender fit each clam with a rubber band to preserve its moisture, and separated them within their cages with impact-absorbing liners. Every action was performed with such a gentle touch that Jeff could easily have been handling ancient Chinese pottery. In contrast, the divers picked as fast as they could. They harvested approximately one cage every ten minutes—which was considered good this time of year at this location. Over the course of the day, the crew gathered and packed thirty-five cages, almost exactly what their buyers ordered.

Due to the uncommon size of the geoducks beneath us, packaging went fast, allowing more time than usual for the crew to reflect on a large part of geoducking—the practical jokes. Lonney spoke about the Good Old Days before the vessel's owner cautioned them to ease up on their extra-curricular activities. But, vendettas have a way of spontaneously erupting

Lonney and Jeff display two very large geoducks.

between the rival crews. Perishable seafood would mysteriously appear in a boat's engine room. Divers would find themselves tied off to a neighbouring vessel. Full clam bags that waited to be brought aboard a boat would disappear in transit. The *Hideaway II*'s owner curtailed his crew's part in these antics when he saw his boat steam into port one evening completely shellacked with broken eggs, another boat having used it for artillery practice.

Before quitting time, the rival boat that had verbally abused us over the radio that morning drifted over to bullshit. The skipper of the *West Wave* looked as though he had ulterior motives for the visit, though his conversation seemed innocuous enough. Said he had reached his quota for this area already, and was going to dive for sea urchins somewhere up the coast. After the conversation the skipper grinned as he fired up the engine and revved it to ear drum-bursting levels as he sped off for the wharf. The gesture was meant for Lonney, who was presently under water, where the sound of a boat half a mile away can seem within a propeller's reach. Later, a completely shell-shocked Lonney got out of his suit and prepared himself a long overdue lunch. He lit the propane burner and looked around cynically, like a school kid who has just received his fourth wedgie. "Okay," he said, looking over to me, "what did they put in here?" No amount of reassurance from me could convince him that his lunch had not been sabotaged. Finally, he poured his leftover chicken cacciatore into a pot and examined it solicitously with a fork. He seemed astonished that it hadn't been tampered with.

Dockside, later that evening all the geoduckers gathered to weigh their catch and load it onto the truck heading to market. Ten to twelve of them stood around talking and jesting with each other. It was then more than ever that this fleet of fishermen resembled a community rather than just a haphazard group of individual fishermen. Their postures suggested tribesmen returned to the cave from a successful hunt; the day was complete. Nearly. It was the jocular skipper of the rival boat who got the last word in: "Yeah, I won't be here tomorrow," he said.

"Fishing for urchins?" asked Dave, taking the bait.

"Yeah," said the other, "so you'll have to follow someone else around." 🐟

THE WRECK OF THE *SALTY ISLE*

Alison Arnot

Higgins Pass

Gale-force winds and twenty foot waves smash the splintering boat against rocks while the crew grab anything on deck and cling for their lives. This image belongs in the climax of an action-adventure movie. But for Ron Bellrose, the scene of the February 18, 1995 wreck of his 73-foot packer *Salty Isle* was very real, although many events that evening were surreal in nature. If it were not for the heroic efforts of several dive boats, specifically three divers who risked their lives to save the *Salty Isle* crew, Bellrose would not be here today to tell his story.

Bellrose, his wife Eileen Hare, and deckhand Mark Barton were heading from Prince Rupert to Laredo Sound, passing the entrance of Higgins Pass when stormy seas overcame the boat. The boat's engineer, Danny Smyrichinsky, had taken the trip off. The *Salty Isle* was packing sea urchins, transporting them from a fleet of dive boats tied up at Grant Anchorage, about 24 miles northwest of Bella Bella.

The *Salty Isle*. Photo: Cathy Converse.

They arrived at Higgins Pass after dark, having been delayed en route by strong winds. Torrential rain was pouring down, causing havoc to the radar screen, and the wind was gusting up to 40 knots. Bellrose says the usual approach to Higgins Pass is to line up the red light on Jaffrey Rock with Kipp Islet, a tiny outcrop of rocks and brush at the entrance to the pass. "But in the rain, we couldn't see the red light, we could see the rock on the radar, but... The plotter said we were right on line, but the [radar] picture wasn't right." Barton lit the front of the boat with the searchlight, and at that moment they could see a light blue patch directly ahead. "Boom we were on the rock," Bellrose says. "Surfing in, in those conditions, there's no way you can stop and back up or move over a little bit. We were doomed. As soon as we hit the first rock I knew we weren't where we were supposed to be. And I knew if we weren't where we were supposed to be—then we were toast in those conditions."

Just after 2000 hours, Bellrose put out a distress call on the working channel of the nearby dive fleet. Then he sent a mayday call to the coast guard. The crew scrambled to get into their survival suits.

Bellrose still tries to put together the pieces of the events leading to the demise of his beloved *Salty*. "What must have happened was, as we were going in, the swell and the surge just moved the boat to the left of the course line. The swells and surge kept lifting us up and moving us in closer on the beach, depositing us. The tide was going out at the same time, so nobody could get near us—and the boat was just getting smashed to pieces."

John Parkin—captain of the *Aquastar*—and a few of the other dive boats anchored a few miles away, responded to the distress call. "We tried a couple of different angles to get at them, but it was too rough. Eventually I realized it was an island and got my way around the back side of it." The *San Juan II* got as close as possible and held a searchlight on the *Salty*. A diver on the *Aquastar*, Greg McKay, went ashore in a drysuit with a rope and hand-held radio, which was smashed against the rocks before he reached the beach. Since they had lost contact, Parkin and Stan Hutchings, a fisheries observer of the sea urchin harvest, headed for shore in a jet boat. Another fisheries observer, Doug Stewart, stayed with the precariously anchored *Aquastar* to make sure she didn't head for the rocks as well.

The three men on land walked about two hundred yards over slippery rocks and brush to the other side of the island. "Waves were just crashing all over the place," Parkin says. "We were only about forty feet away from Ron—everything was just a mess. We couldn't hear anything he was saying because there was such an incredible noise coming from the wreck. The only sounds were timber smashing and wind blowing." Despite the conditions, the men on shore managed to motion to Bellrose to throw something to them. Or perhaps they both had the same idea at the same time.

"I figured if I could get a rope to these guys on the beach, we could go across, down the line," Bellrose says. He tied a life-ring to a rope, went to the bow and drifted the line in through the surf to the men on shore.

But this process was not as easy as it sounds. It took about an hour to get the line secured from the boat to shore, Bellrose says.

"We couldn't get [the rope] because of the way the wind was blowing," Parkin says.

Bellrose expressed admiration for the efforts of the three men on shore. "I could see these guys trying to get at this life-ring and just being completely overpowered in the surf, bounced around and up on the rocks. These guys were actually risking their lives." Finally the rope drifted close enough between the rocks and swells for Parkin to jump down and grab.

In the meantime, the *Salty Isle* crew had to deal with the boat disintegrating around them. The foremast had fallen across the deck, forcing Bellrose to crawl underneath and around all the rigging that had come loose. Giant waves were crashing over the top of the boat. Bellrose says he could hear when a big wave was about to thunder upon him, at which point he had to stop whatever he was doing, grab onto something, and flap like a flag in the wind until the water washed away again.

They had intended to jump the 15-foot drop to the water and crawl along the line, but just as they were about to abandon ship, the chainplates supporting the

A dive boat alongside the *Salty Isle*. Photo: Cathy Converse.

main mast started to lift. "I knew the main mast was going to go," Bellrose recalls. "I tried to push them [Hare and Barton] backwards to go the other way. We couldn't talk, they couldn't hear what I was saying—so I heaved them backwards." The main mast went crashing across the wheelhouse, leaving the crew clinging to the bulwarks as the boat tore apart. "Eventually one big giant wave hit us and the whole boat went over sideways. The wheelhouse lifted right up, all three of us were torn off the bulwarks and swept underneath into the guts of the ship," Bellrose says. "We all figured we were doomed at that point. We just thought the wheelhouse would slam back down and that would be the end of us. But it didn't come back down. It went right off the boat."

The same wave that spelled doom for the *Salty Isle* also created havoc on shore. It yanked the rope right out of McKay's hand. Parkin recalls, "The mast snapped and the wheelhouse got knocked off and we lost sight of them. We thought it was over." Parkin says that during the ten minutes when the *Salty Isle* crew had disappeared, the men on shore were expecting to see bodies surface.

But, "eventually we bobbed to the surface," Bellrose says, feigning nonchalance. "I had my foot caught for awhile. It's a bit freaky under water with your foot caught."

Parkin could not hide his disbelief concerning the final events of the rescue. "With my flashlight shining on the boat, I moved a little closer and there they were. They were all huddled on the deck, on the bulwarks. And then, like a miracle, the rope drifts right in front of me." Parkin grabbed the rope once more and yelled for them to jump in the water, which they immediately did, grabbing the rope at the same time.

The wreck victims pulled themselves along the rope to the shore a few yards away, where the others hauled them onto the rocks. "If they hadn't been there, after all we had gone through I don't even know if we could have climbed onto the rocks. We were just exhausted at that point." Hare was so exhausted and distraught she could not walk. Hutchings had to carry her the two hundred yards to the jet boat on the other side of the island, which Parkin says was a feat in itself considering how slippery the rocks were. On returning to the island in daylight, he says the rocks and brush were so precarious he could barely walk on it then. "I guess it was a miracle nobody broke a leg because the rocks were so slippery, covered with diesel fuel and stuff."

Bellrose adds, "With such a huge ship and so much violence happening for such an extended period, it's just a miracle no one got hurt."

At about 2230 hours, the survivors of the shipwreck were taken aboard the *Pub Nico Gemini* for a change of clothing, showers and much needed cigarettes. The Coast Guard boat *Arctic Ivik* arrived on the scene about 2300 hours.

All involved believe forces beyond their control were at work that night. "It was incredible. It was too spooky," Parkin says. "It was like that boat didn't want to take anybody with her. It was her time to go and she didn't want to hurt anybody. We really didn't do anything. It was sort of like everything was supposed to happen that way."

Hare, her voice trembling, echoes these sentiments. "We really felt that the boat didn't want to hurt us. She really was a live person. She was going and she didn't want us."

Ironically, the *Salty Isle* went down on the same day as the memorial service for the man who named her. Daryl Georgeson, an old towboater living on Salt Spring Island, had won the contest held in 1984 by her owner at the time, Hirk Roland, to rename the boat he had just bought. Since her arrival in Canada in 1976, the boat had been called *Choice One*. Previous to that, she had fished the North Sea under the name *Fertile Vale*. She was built in 1954 by Richard Irvin & Sons Ltd. in Peterhead, Scotland. Bellrose had owned her for about a year and a half after working several years for Roland.

Parkin and McKay returned to the wreck site two days later, looking for Hare's bracelet among the refuse now littering the seabed. They didn't find it, but they did find an unbroken bottle of Cabernet underneath a chunk of ballast. "We couldn't believe it. That was the only thing that wasn't broken in the whole wreck." They retrieved the wine and gave it to Bellrose and Hare, who plan to use it to christen their next boat.

AFTERWORD

THE OLD CAPTAIN TAKES A WALK IN THE SUNSHINE

Mrs. Amor de Cosmos

I watched him as he walked along the path to the parking lot at the Steveston fisherman's dock. I knew he was making his way to the dock when I spotted him earlier, over at the other dock where they sell fish and shrimps to tourists for cash.

I also knew who he was. He was an old fishing captain taking a walk in the spring sunshine. But it was not just any old walk. No, he was doing something that comes to us all: satisfying the need to relive the past through looking at familiar and unfamiliar things. He walked to look at all the boats tied up in Steveston, and to let his memory do its work. Sort of like reading a book many years later and finding things you missed the first time around, but also seeing old things again.

I had time to examine him closely as he walked to the dock. He wore a black suit of heavy cloth that looked like he had bought it long ago only to wear on serious occasions. His shirt was white and his tie was so old it was about to become fashionable again. He wore sensible black shoes and though I couldn't see from where I was, I knew the socks were hand-knitted, and probably hand-mended. I was sure a wife or someone from the old days had made them. I even knew the label of the hat he wore. It was a brown fedora that went out of style with gangster movies, but it was not just any old hat. No, it was a Borsalino, and I recognized it. The old captain was presenting himself to the world and he had to do it just right.

The hat was the key, for I knew him even though he did not know me. When I was a kid on the Island, this man was feared and respected. He was a top fisherman; what we call a "highliner." The hat was given to him by a fishing company at the annual Christmas party, recognizing him as the top seine boat captain on the Island. They don't do things like that any more. Anyways, you can't wear a fedora to Hawaii or Palm Springs.

As the old captain walked by me I tried to figure out his age. He was over seventy and life had not been kind to his body. Arthritis in his hands from longlining, bad knees, and probably high blood pressure from the fisherman's diet of cigarettes and coffee.

But his eyes were not old. He seemed to have two ways of looking. One was the normal way, the other was what I call the "zoom mode" where he could be looking at you and talking, but his eyes were scanning the horizon and checking every detail of boat rigging and how the water was and the state of the tide. A million details in a glance—that came from fishing in the days before radar and sounders.

By now he was down on one of the floats looking at the boats tied there. The aluminum super skiffs he barely acknowledged. Nothing like them when he fished. Same with the fibreglass gillnetters. But when he came to an old wooden boat, he paused and took a long hard look.

I am sure he knew many of the boats well. One thing about the BC fishery is that its boats last a long time. Many of them are over fifty years old. So the old captain could look at a boat and think of many things. Hydraulics. The transition to drum seining. Nylon replacing manila and cotton, forcing fishermen to learn new knots. But it was always the fish that he came back to. Now I don't really know that, but I was trying to put myself in the old captain's place. I mean, I could have gone over and talked to him and asked him point-blank what he was thinking about, but I didn't. Somehow I thought it was private.

But what fish he must have seen! The pink run in 1962, when the fish came to the Namu area for months. The '58 sockeye year, when they came through the Straits. And dog salmon fishing for weeks on end—a season that lasted from June to November. Fish and places we would never know. They didn't have the power and the gear we have now. Sets were planned and carefully executed, not just banged out like now. Of course, the good old days usually weren't. No Workers' Compensation, no Unemployment Insurance, and no pension plan. Prices were so low that fish were counted as pieces, not by weight like today.

My good old days are right now, and I want to make the most of them. That is why I watched the old captain with such care. One day I will be like him. I will come to the Steveston dock and look at the boats and dream of fish and places. I will look for my past in the boats, for fishermen grow nothing, build nothing, leave nothing behind. All we can do is look at a boat and capture what it was like for us. All of us have to take a walk in the sunshine, we can only hope that it will be pleasant. 🐟

Overleaf: The FV *Caledonian* in a November gale in Hecate Strait, 1989. Photo: Peter A. Robson.

INDEX